IEE MONOGRAPH SERIES 14

Electrical insulating materials
and their application

IEE MONOGRAPH SERIES 14

Electrical
insulating materials
and their application

R. W. Sillars, B.A., D.Phil., F.Inst.P., C.Eng., F.I.E.E.
Manager, Trafford Park Laboratories,
GEC Power Engineering Ltd.,
Manchester M17 1PR, England

738 . 131

Peter Peregrinus Ltd.
on behalf of the
Institution of Electrical Engineers

Published by Peter Peregrinus Ltd.
Southgate House,
Stevenage, Herts. SG1 1HQ, England

© 1973: Institution of Electrical Engineers

ISBN: 0 901223 40 9

Printed in England by
Billing & Sons Ltd., Guildford and London

No longer can one person write authoritatively on every aspect of electrical insulation, but neither can a book of reasonable size cover fully all the ideas and information that are useful. What can be done is to give a guide to the rapidly accumulating knowledge of increasing numbers of materials, and to the steadily growing framework of theoretical concepts by which this knowledge can, in some degree, be organised and rationalised. It seemed to me that such a guide could as well be written by one author as by many, with perhaps some gain in compactness and unity.

I have tried in the first, background, chapters to show how few basic theoretical concepts are really necessary to illuminate the mechanical, thermal and electrical behaviour of polymers; I have presented these with no rigour whatever, though indicating, I hope, where the gaps in the presentation lie and the sources from which they can be filled.

The vital subject of testing cannot be usefully dealt with in terms of general principles, while summaries of procedures (which in many cases are arbitrary) are both dangerous and tiresome. I have therefore tried the method of indicating the nature of each test and leaving the reader to consult for himself the authorities quoted, where he wants full information. In the present period of transition between national and international testing methods it seemed worth while to quote some representative national standards as well as the growing body of international ones; US and German standards have therefore been quoted with British standards.

The accounts of materials themselves, in Chapters 7–11, cannot be entirely forced into a common pattern, but the plan has been to include the following aspects, generally in the order given here, except that some numerical values are tabulated: sources or methods of synthesis, chemical nature, types and modifications, methods of processing, density, mechanical strength, rigidity and hardness, temperature limitations, susceptibility to moisture, susceptibility to solvents, permittivity and losses, susceptibility to discharge and tracking, principal uses and reasons for suitability for these. I have also tried to cover the important but difficult question of relative cost.

It is a great pleasure to be able to thank many friends and colleagues for help, direct and indirect, in connection with this work. There are the many whose collective experience, imparted over several years of discussion of insulation matters, has gone into what is written here; particularly there are those who have spent time and effort reading drafts of this monograph and giving advice: E. Farmer, T. Holt, J. S. Simons, L. M. Sloman, R. Snadow and F. B. Waddington. I am also very glad to thank the authors who are acknowledged on various illustrations for permission to use their results, especially W. Reddish and ICI Ltd. from whom I have had valuable advice and who have gone to particular trouble to prepare Figs. 20 and 21. I am likewise indebted to the following publishers for permission to reproduce illustrations:

Faraday Society (Figs. 17, 20, 21); British Association for the Advancement of Science (Fig. 27); John Wiley (Figs. 1, 2, 36, 41, 42, part of 43); International Union of Pure & Applied Chemistry (Fig. 42, part of 43); Conférence Internationale des Grands Résaux Électriques a Haute Tension (Fig. 50); Butterworth (Fig. 46); Institute of Electrical & Electronics Engineers (Figs. 48, part of .49); Electrical Research Association (part of Fig. 49); Royal Society (Table 5); American Chemical Society (Table 6).

Most particular thanks are due to Miss B. Claxton for her patience and perseverance in producing the typescript despite my innumerable alterations and her many other tasks. Finally I thank GEC Power Engineering Ltd. for permission to publish this work and for the use of their facilities during its preparation.

R. W. SILLARS

Contents

viii   Contents

4.2.3.1   Heterogeneity   55
4.2.3.2   Ions with limited travel   57
4.2.3.3   Electrode and barrier effects with ionic   59
          conductivity

4.3   Electric breakdown of solids   60
    4.3.1   Discharge-free ('intrinsic') breakdown   62
    4.3.2   Progressive breakdown in solids   63
    4.3.3   Breakdown initiated by extraneous discharges
            (partial discharge or 'corona' breakdown) 65
    4.3.4   Breakdown in electric strength tests   68
    4.3.5   Internal discharges, their assessment and
            prevention   68
    4.3.6   Thermal breakdown   71

4.4   Electric breakdown of liquids   73

4.5   Tracking and tree-burning   75

5   Moisture in insulation   80

5.1   Effects of moisture   80

5.2   Removal of moisture from high voltage insulation 86
    5.2.1   Transformer drying   87
    5.2.2   Cable drying   88

6   Test methods   89

6.1   Mechanical properties   90
    6.1.1   Tensile properties (short time)   90
    6.1.2   Flexure properties and cross-breaking strength 91
    6.1.3   Impact properties   91
    6.1.4   Compression properties   91
    6.1.5   Tensile creep and stress-relaxation   92
    6.1.6   Hardness and indentation   92
    6.1.7   Tear resistance   93
    6.1.8   Abrasion resistance   93

6.2   Thermal limitations   94
    6.2.1.   Temperatures of deformation or softening 94
        6.2.1.1   Heat deflection (or distortion)
                  temperature   94
        6.2.1.2   Cantilever deformation tests   94
        6.2.1.3   Vicat softening point   95

A*

Chapter 1

Introduction

Many branches of technology develop empirically, independently of any exact science which may bear on them, others were born of an exact science and can only be developed by a knowledge of this science; a few (of which semiconductor technology is a notable one) have progressed from the first kind to the second. The technology of electrical insulation is likely to be in the first of these categories for many years, but doubtless it will at some time pass into the second. The sciences which bear on it are polymer chemistry, solid state physics and that branch of solid state physics which deals with dielectrics; but the insulation technologist soon realises that the practical help he can get from these is limited, and differs according to the type of electrical equipment he is dealing with.

This monograph is a review of the present state of insulation technology addressed to the engineer rather than the scientist. It assumes the reader is familiar with the terminology used by electrical engineers especially in relation to material properties, but explanations of terms and methods of measurement where desirable are given in Chapter 6. It also assumes he has a knowledge of school chemistry (he can get by without this), and of physics to first year university standard.

It is not intended as a comprehensive text: the long-established topics such as thermal classification, dielectric breakdown, and the properties and uses of the more traditional materials are given less space, relative to newer topics, than their practical importance warrants; the reader who wishes to become fully informed will have to read some of the older textbooks referred to, the contents of which are not reproduced here.

Chapters 2 and 4 will be concerned with some aspects of the sciences of polymers and dielectrics because, despite their limited usefulness, they help to co-ordinate the subject into a pattern which can be comprehended; and because the insulation engineer communicating with scientists trying to improve existing materials needs some degree of common language. It is to be hoped that some technologists will not be permanently content with this small degree of overlap with their scientific colleagues. Chapter 4 has been included to indicate the directions in which an understanding of dielectric behaviour is now developing.

Occasionally a knowledge of scientific principles may give an insight (real or imaginary) into a practical problem, but whatever effort he may make to use the sciences the insulation man finds himself in a largely empirical technology. His problems are those of meeting the limitations imposed by cost, design of the equipment to be made, types of manufacturing equipment at his disposal and the materials available on the market; his guide lines are current practices and his main sources of information are articles in technical and trade journals, and communications with suppliers and with colleagues.

The published values of electrical properties are frequently unhelpful in choosing materials for particular applications. This is especially true of breakdown strength and conductivity, for two reasons: firstly, if the final symptom of failure of insulation is electrical breakdown, this usually results from some progressive process such as embrittlement by heat and/or oxidation to which electrical breakdown strengths give no clue; secondly, these properties are often determined by impurities or defects which may be characteristic of a particular condition, or source, or method of manufacture.

When he wishes to depart from current practice, whether to improve an insulation system, or to overcome a manufacturing defect, or to reduce overall cost, the insulation man must cast about for materials and processes which may fit his need. Unless the change is a small one he must devise simulation tests to see whether the new materials meet all the service conditions; he may have to do extended life tests and to satisfy himself that any 'acceleration' of the tests is valid. (Chapter 3 and Section 4.3.3 outline the basic principles on which most of such tests are based.)

He must be sure that the materials he tests are going to be reproducible and that they have been sufficiently characterised to enable him to verify in the future that he is using the same ones. He must also devise checks on the processing techniques so that if these were changed inadvertently, or in ignorance, he would find out. A great many of the tests to which materials are subjected are simply intended for these purposes. Tests of viscosity of a varnish or loss angle of a laminate, for example, are not necessarily dictated by some unique viscosity required to make the process workable, or by some limit of loss angle imposed by the application; they may be simply methods of checking consistency of manufacture.

Chapter 6 reviews the methods of test used to characterise materials and to gauge their suitability for particular applications. Simple tests which have been in use for many years are covered briefly with references to national or international standards; those which have appeared or changed in the last ten or fifteen years, especially those based on simulation of service conditions, are discussed more fully.

The importance of moisture in insulation and its removal is so general that it is convenient to include Chapter 5 on this topic rather than deal with it piecemeal under the various materials which are susceptible to its effects.

Individual materials, their sources or methods of synthesis, chemical nature, methods of processing, properties, limitations and principal uses are dealt with in Chapters 7–11. The better-known materials, especially those which have been largely covered in the older textbooks referred to, are dealt with sketchily; the more recent synthetic polymers are considered as fully as space allows, with references to recent literature.

Chapter 2

Broad structural types
and their mechanical behaviour

2.1 Relevance of primary chemical structure

In organising one's thinking about materials, it is useful to recall the different ways in which atoms, ions and molecules are bound together in a solid, packed as close to each other as their 'size' (a rough but useful concept) permits. The binding forces may allow minor vibrations only, or may give freedom for one group of atoms to rotate relative to another group (and therefore for chains of atoms to bend and contort), or may permit groups of atoms or molecules to slide or jostle bodily past one another (as in a liquid or plastically deformed solid) without complete separation. Different degrees of freedom prevail in the same material at different temperatures, even though there may be no major change such as melting. The differing mechanical properties of insulating solids can be better understood by appreciating the different combinations of inter-atomic and inter-molecular forces operating.

The strongest bonds between atoms, the 'chemical' bonds, arise in general from shared electrons. In terms of the crude picture that a single atom consists of electrons—point negative charges—orbiting round a positive nucleus, a molecule consisting of two atoms has one or more electrons in outer orbits which encircle both nuclei. We shall not try to provide further insight into the nature of chemical bonds, the concepts involved would take too much explaining; we shall simply consider some consequences of different types of chemical bond and of combinations of them.

At one extreme is the pure ionic bond where the 'sharing' takes the extreme form that all atoms of one type each donate an electron to atoms of another type, so that each resulting ion electrostatically attracts all oppositely charged ions and repels all similarly charged ones. This results in a regular crystalline solid in which positive and negative alternate. The classical example of this is sodium chloride, a cubic structure in which each ion is attracted equally to six oppositely charged nearest neighbours. This structure is illustrated in Fig. 1 (Plate); it is introduced merely as the simplest and most easily visualised archetype of a structure in which there are no molecules; every unit is equally bonded to several others so that relative movement is impossible without dissolution or rupture of the structure. (Movements associated with structural defects are ignored for simplicity at this stage.) In fact all purely ionic crystals are water soluble and of little practical interest as insulation, but there are many insoluble substances, notably ceramics and glasses, in which each atom or ion is attracted or attached more or less equally to several neighbours by a combination of electrostatic attraction and complex electron sharing in which the 'orbits' of the electrons embrace groups of neighbouring atoms. These are more difficult to visualise and explain, and they are not necessarily crystalline, but they have the characteristics of rigidity (high elastic modulus), often combined with high mechanical strength, hardness, and brittleness when distorted beyond a certain limit. These ceramics and glasses have as their main constituents metals such as magnesium, aluminium, silicon, calcium, titanium, boron (many others are possible) together with oxygen. Thus solids such as magnesium silicates, aluminium silicates and magnesium titanates have no separable molecules; any chemical formula that may be written for them represents a repeating unit in an extended structure rather than an independent entity such as a water or oil molecule. The substances just mentioned are rigid and also crystalline, though the crystallinity is not usually recognisable by casual inspection; they consist of microcrystals each perhaps a micrometre or less in diameter, without the obvious facets we associate with crystals, but showing an X-ray diffraction pattern which indicates regular and accurately maintained distances between similar planes of atoms. There is not necessarily a space between the boundaries of the minute crystals and the only discontinuity may be a change of orientation and perhaps a certain amount of misfit between the crystal units on opposite side of the boundary surface. Fig. 2 (Plate) illustrates a structure which is typical of the materials formed from fired clays in porcelain manufacture.

There are a number of less well-defined substances made up of some of the same elements which are rigid and non-crystalline, i.e. they are glasses. Each atom is still bonded to several neighbours but there is no orderly and precise repetition of structure even on an atomic scale; X-ray diffraction shows merely that certain roughly maintained spacings and groupings are common through-out the structure.

In contrast to these structures, in which each atom (ion) is bonded to several near neighbours, is the 'chain' structure in which the atoms form self-contained small groups or units, and each unit is bonded only to two other neighbouring units, so that if one were to move from unit to unit along the bonds there would be no choice of path, one could only move along the chain. The commonest chain molecules have a 'spine' of carbon atoms.

Carbon normally forms electron-sharing bonds with precisely four neighbouring atoms, the bonds being symmetrically directed outwards, i.e. directed to the four corners of an imaginary tetra-hedron surrounding the atom. If all the other four are other carbon atoms, and these four are linked to still other carbon atoms, we again have a rigid structure, the diamond crystal. If, however, two bonds of a carbon atom are linked to hydrogen atoms, two bonds remain which can connect to other similar units of carbon atoms each with two hydrogens, and so on, we then have a chain molecule and no chemical bonds are left to form links outside the chain (Fig. 3*a*) (Plate). A solid composed of such linear chains side by side and free to slide past one another can obviously be bent or subjected to various distortions without rupture. These 'molecular' chains are frequently tortuous and intertwined so that there is no regular spacing of planes and therefore no sharp X-ray diffraction pattern. However, it is possible for hydrocarbon chains to be straight as in the model of Fig. 3*a* (disregarding the zigzag imposed by 120° angle between carbon–carbon bonds) and a bundle of such straight chains is regular and crystalline. Paraffin wax, for example, consisting of chains about 20 carbon atoms long, crystallises readily. Poly-ethylenes, consisting of chains of the same units but hundreds or thousands of carbon atoms long are unlikely to form single crystals, but have crystalline regions in which parts of different neighbouring chains (or parts of the same chain doubled back on itself) form into regular bundles. These unbranched and lightly branched chain polymers may be amorphous, partly crystalline or very rarely) wholly crystalline.

A carbon atom of one of these chains, instead of having two hydrogen atoms attached to it, may have one hydrogen and another carbon atom, forming a branch (see Fig. 3*b*). This branch may be a single carbon atom with its remaining bonds satisfied by three hydrogens, or it may be another chain similar to the main

one. Such branches will clearly get involved with neighbouring molecules and if they are frequent they will tend to restrict the relative movements and make the solid more rigid. They also tend to prevent crystallisation occurring.

In some polymers the branches may in fact be, or connect to, other major chains, so that all chains are linked to other chains, this is known as a cross-linked structure. The kind of molecules that we have been talking about so far, consisting only of carbon and hydrogen, cannot be cross-linked to a very great degree; but some molecules involving groups containing oxygen and nitrogen can be very highly cross-linked, and such materials are then rigid and infusible, like the phenol formaldehyde ('bakelite') resins.

Solids composed of long chains of atoms, whether unbranched, branched or cross-linked are called *polymers*, usually with the implication that the chains are hundreds or thousands of units long. The individual representative units of the chain are called *monomer units* and the substance consisting of separate monomer units, if it exists, is called the *monomer*. Thus ethylene is the monomer of polyethylene, and so on. The reader probably knows that carbon atoms form other kinds of bonds than the fourfold symmetrically directed ones we have considered hitherto; notably the double bond and the benzene ring. Many monomers contain one or more

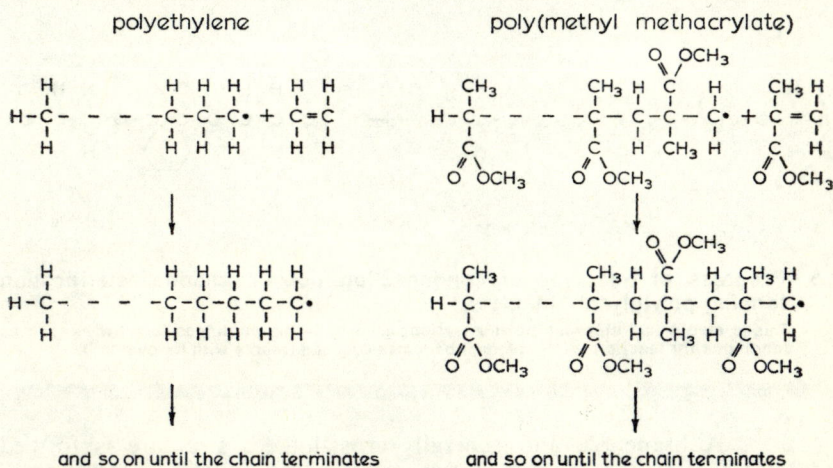

polyethylene   poly(methyl methacrylate)

and so on until the chain terminates   and so on until the chain terminates

Fig. 4 Progress of addition polymerisation of polyethylene and poly(methyl methacrylate)

Starting from an activated radical the molecule grows rapidly until the activation ceases; it does not start to grow again thereafter

double bonds, as does ethylene, linking its two carbon atoms. One type of polymerisation, 'addition polymerisation', can be regarded as the 'opening' of one of these double bonds and attachment of one of the 'loose ends' to the 'loose end' of a growing polymer chain. Fig. 4 shows this process diagrammatically, without indicating the mechanisms by which it can take place.

Polymers are not necessarily simple carbon–carbon chains of exactly the sort which we have hitherto concentrated on for simplicity. The backbone may involve oxygen atoms, nitrogen atoms, benzene rings, silicon–oxygen–silicon chains (silicones) and other features. Many of them are formed not by double-bond splitting (addition) but by a reaction such as esterification (reaction of an organic acid or anhydride with an alcohol) with the elimination of water (condensation). An unbranched chain is formed when an alcohol with two OH groups (i.e. dihydric) such as glycol reacts with an acid having two COOH groups (i.e. dibasic), such as phthalic acid (Fig. 5).

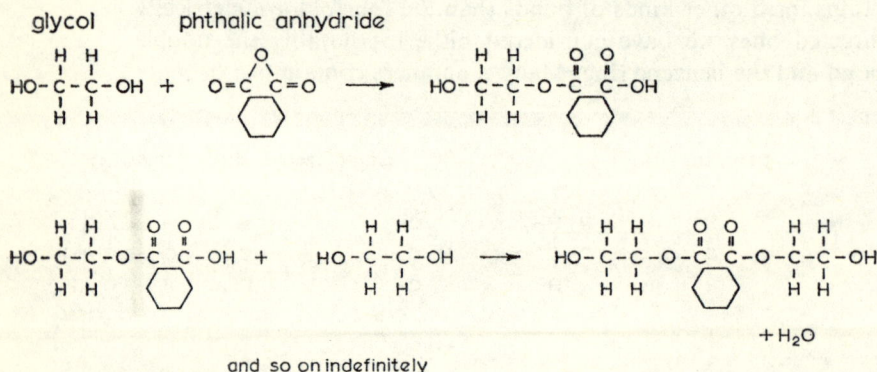

Fig. 5   Progress of one type of condensation polymerisation: esterification forming poly(glycol phthalate)

This is also called stepwise polymerisation; each molecule continues to grow as long as conditions for reaction exist, unless it becomes cyclised (reacts with its own tail)

A branched, and generally cross-linked, structure is formed when, for instance, a trihydric alcohol such as glycerol is involved; each glycerol molecule can form three ester links with say, phthalic acid, which can then link to other glycol or glycerol units and so on. Polymers formed by esterification are generally called polyesters; but polymers in which the ester group is not part of the

IMPORTANT NOTICE TO READERS

Have you, or has your library, considered the possibility of taking out a **Standing Order** for our publications in this field? Under this scheme, you would receive all new volumes, either in our various series or over our complete publishing programme, as you wish, as soon as they become available. Should you find at a later date that the volumes no longer serve your needs, you may cancel your standing order without notice.

Please write for a copy of our latest catalogue, which gives details of books and periodicals published by Peter Peregrinus Ltd.

The Publisher will be pleased to receive suggestions for consideration for future publications.

backbone but incidental, as it were, to a double-bond polymerisation, such as poly(methyl methacrylate), may also be referred to as polyesters. There are many less direct ways of bringing about the condensation process than that indicated in Fig. 5.

We have now recognised four types of insulating solid.

(*a*) Rigid microcrystalline inorganic solids such as alumina and low-loss ceramics made up of accurately repeated structural units throughout each microcrystal in all directions, in which each atom/ion has bonds to several neighbours. These generally have densities in the range 2·4–4·0 and elastic moduli of the order of $10^5 \, N/mm^2$.

(*b*) Rigid amorphous inorganic solids—glasses—in which roughly repeated groups of atoms recur in all directions, with spacings which are variable, but in which each atom/ion has bonds to several near neighbours. Densities and elastic moduli are similar to those of type (*a*).

(*c*) Molecular solids, generally organic, in which a unit structure recurring at accurately spaced intervals along a chain of primary bonds, such as illustrated in Fig. 3, forms the molecules which may be assembled:

   (i) in straight and regular bundles of chains (cf Fig. 3*a*), with accurately repeated spacings in all directions (crystalline polymer);

  (ii) tangled together with no apparent regularity (amorphous polymer);

 (iii) in a mixture of regions of (i) and (ii); a crystalline region often including several lengths of the same chains (folded back and forth) in a bundle;

 (iv) like (ii) or (iii) but with occasional cross links or 'bridges' of primary bonds between chains. These, under certain conditions, behave as elastomers (like natural rubber) and are discussed in Section 2.4. The densities of types (*c*. i) to (*c*. iv) are generally in the range 0·9–1·5, the lower values relating to those which contain carbon and hydrogen only. They are much less rigid than the inorganic solids, generally having elastic moduli in the range 200–4000 N/mm^2, the lower part of the range applying to polythene and similar flexible materials. Type (*c*. iv) materials in the elastomeric condition (which occurs at temperatures where, but for the cross links, the polymer would be a viscous liquid or plastically deformable solid) have elasticities in a much lower range, e.g. for soft vulcanised natural rubber around $10–100\,N/mm^2$, and for uncured natural rubber as low as $1\,N/mm^2$ (but in this condition unable to support a sustained load). Type (*c*) materials are generally called *thermoplastics*.

(*d*) Quasi-molecular organic solids, made up of repeated groups of atoms, some or all of which are connected to three other groups forming a cross-linked or 'three-dimensional' structure, which is typically more rigid than that of type (*c*) but less so than the inorganic solids, types (*a*) and (*b*). Densities are in the range $1\cdot3–1\cdot7$ and elasticities $1000–5000\,N/mm^2$, but these materials are normally used with fillers which increase both figures considerably. Type D materials are generally called *thermosetting*, because they are usually produced by curing at a fairly high temperature and, once cured, cannot be softened by heating a second time.

For the remainder of this chapter we shall consider the polymeric types (*c*) and (*d*) only.

2.2 Forces between molecules — secondary bonds

The bonds we have been discussing so far are the strong 'primary' or 'chemical' bonds; we have seen that in most hard solids such as diamond, porcelain and glass every atom has strong bonds to several neighbouring atoms, and in the cross-linked organic resins of intermediate hardness, such as phenol formaldehydes, there is a three-dimensional network of strong bonds, such that a substantial proportion of units are each linked to at least three others.

In contrast to these, substances such as wax, sugar and bitumen consist of molecules of limited size held to one another by 'weak' or secondary forces. The much larger long-chain molecules of the polymers of type (*c*) are also held together by secondary forces whose existence we have, so far, tacitly assumed. These are of various kinds, known collectively as Van der Waals forces, because they correspond with the term for attraction between molecules introduced by Van der Waals as a correction to Boyle's law.

Before discussing the nature of these forces we must look more closely at variants of the electron-sharing (covalent) bond. When carbon atoms are linked only to hydrogen and to other carbon

Fig. 6 Diagrammatic representation of charge distribution in dipolar molecule methyl chloride

atoms there is a high degree of symmetry in the distribution of the electron 'orbits' and therefore in the average electric charge density. If, however, one of the bonds is between carbon and chlorine, say, as in methyl chloride, the chlorine takes a share in an electron of the carbon atom but does not itself contribute one, so that there is a preponderance of negative charge at the chlorine atom and a deficit of negative charge, i.e. a positive charge at the carbon (Fig. 6). This, then, is a permanent electric dipole, and a molecule such as methyl chloride is called a dipole molecule. (Frequently it is called a polar molecule, but the word polar has so many different connotations—it is also used to describe an ionic crystal—that it will be avoided here.) Likewise any group in a molecule having asymmetry of charge, such as those of Fig. 7, is called a dipole group. It will be obvious that dipole groups attract each other if suitably oriented, and also induce temporary dipoles of opposite orientation in neutral molecules or groups (just as a magnet induces magnetisation in a piece of soft iron), so as to attract the neutral molecule to the dipole. Note that the bond in a

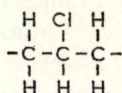

chloride
1·9 – 2·1

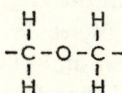

ether
1·2

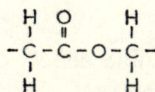

ester (main chain)
1·7 – 1·9

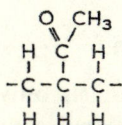

ester (side group)
1·7 – 1·9

Fig. 7 Examples of dipole groups common in polymers

Figures indicate the dipole moment in Debye units (1 Debye unit = 10^{-18} ES units = 3.33×10^{-30} Cm)

dipole group is a *shared* electron; unlike the sodium chloride crystal where the metal loses its electron completely to a chlorine and attracts all neighbouring chlorine ions equally, the dipolar bond is confined to a particular pair or small group of atoms.

The idea of the attraction of a permanent dipole molecule of this kind for a neutral molecule is a convenient introduction to the simplest way of explaining the feebler attraction between non-dipole molecules. Although a group such as —CH_2— has no permanent electrical asymmetry, thermal agitation slightly distorts it, causing fluctuating dipole moments which attract similar neighbouring units, and vice versa. This attraction is known some-what obscurely as 'dispersion' or 'exchange' attraction (or it is sometimes called after London who provided the first quantitative theory of it).

A fourth type of inter-molecular attraction is generally known as 'hydrogen bonding'. A hydrogen atom belonging to one molecule, situated near to an electronegative atom (i.e. one which readily accepts an additional electron, such as chlorine, oxygen or nitrogen) belonging to another molecule, will tend to establish a feeble electron-sharing or exchange partnership. This type of bond is usually effective between an oxygen, nitrogen or fluorine atom, and the hydrogen of a hydroxyl, carboxyl, amine or amide group.

Now our picture of a linear polymer is of a number of chains, sometimes side by side but more often tangled up, each chain formed by primary bonds which for the present purpose are regarded as unbreakable, the chains being held to one another by the forces we have described as secondary, acting between con-tiguous units belonging to neighbouring chains.

It is reasonable to expect that the greater the number of units in a chain the more strongly the molecules are bound to each other, and this proves true. Moreover, molecules, or parts of molecules, which lie in a crystalline—and therefore compact—configuration, are more strongly bound to each other than those in an amorphous region.

2.3 Physical effects of temperature

In this section we shall discuss only the reversible effects of temperature, excluding oxidation, decomposition, etc.

We have been speaking of strong or weak bonds without saying how such strengths are measured. Although the concept of a 'force' necessary to break a bond is the obvious interpretation of the word 'strength', such a quantity is not directly measurable, and the quantitative measure we have in mind is the energy, or work, required to separate the bonded atoms; this is generally measured in terms of the temperature at which the bond ceases to be effective. The concept usually invoked is of a trough of potential energy (plotted against distance between the atom centres) surrounded by a crest (Fig. 8). The bottom of the trough is the equilibrium position between the attraction forming the bond, and the repulsion which arises when the outer electron orbits (other than shared ones) are very close (crudely, when the atoms bump into each other). The detailed interpretation of such diagrams is complex, the point to be noted here is that a definite energy labelled W in the diagram, must be supplied to the bond to break it.

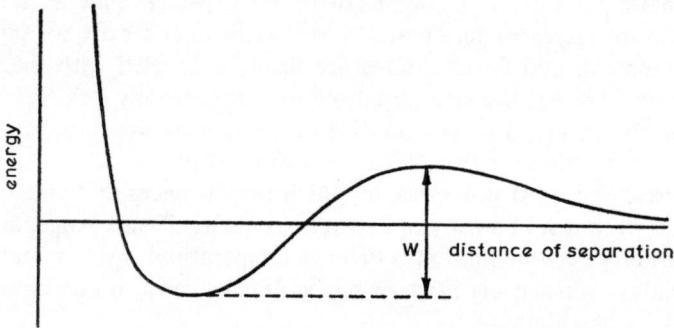

Fig. 8 Cross-section of energy diagram for separation of two atoms bound by secondary forces, with energy W required to separate them

The random interchange of thermal energy between elements (atoms, ions, groups) results in a distribution of energy which (in circumstances such as we are considering where quantum considerations can be ignored) is roughly Gaussian. The most probable translational energy of one atom is kT where k is Boltzmann's constant and T is the temperature in kelvin; the fraction of the total number of units which can escape over an energy barrier of height W is approximately proportional to

$$\exp\left(-\frac{W}{kT}\right)$$

Thus, for example, the rate of evaporation of molecules from a

liquid surface in a vacuum, proportional to the number of molecules which acquire the energy necessary to escape from the
attraction of their neighbours, varies with temperature in this way.
The number of units which, at a given moment, are free of the
attraction between dipoles, or of a hydrogen bond, is governed by
a similar relationship; moreover, the probability that a bulky
molecule or chemical group, packed among similar molecules or
groups, happens to be sufficiently free to slip past one of its
neighbours can be thought of in the same way. This roughly
accounts for the near-exponential decrease of viscosity with
increasing temperatures commonly found in viscous liquids.
(There are more precise and refined descriptions of viscosity/
temperature relationship but the most important feature is the
roughly exponential relationship.)

Not only in liquids but in amorphous solids such as a glass or a
long-chain polymer, it is therefore to be expected that as the
temperature increases fewer and fewer weak bonds exist at any
given moment, and fewer groups are firmly entangled with their
neighbours, so that the structure becomes more readily deformed.
Conversely, at very low temperatures all polymers are hard, and
may be so brittle that they cannot be successfully used.

Glasses, and most polymers, in which the distances and attractions between neighbouring units vary over a significant range, do
not abruptly become fluid at a definite temperature as do crystalline solids in which every atom or molecule is in identical surroundings to its neighbours.

We have to distinguish between softening temperature, which is
somewhat arbitrarily defined in terms of mechanical stability, and
crystalline melting point which is defined as the temperature at
which X-ray diffraction ceases to show any crystalline order. The
latter temperature is usually quite well defined.

The softening temperatures of polymers are raised by:

(a) chain length, the longer the chain the more secondary
     inter-chain bonds have to be freed simultaneously to
     allow independence;

(b) stronger secondary bonds between groups in neighbour-
     ing chains; the polyamides have higher softening
     temperatures than non-polar hydrocarbons of similar
     chain length;

(c) crystallinity, which brings neighbouring chains closer
     together with their monomer units more completely
     fitted to each other;

(*d*) rigidity of the chain in the molten state; e.g. polytetrafluorethylene (rigid because of the size of the fluorine
atoms) has a much higher softening temperature than
polyethylene.

The crystalline melting point is similarly affected by (*b*) and (*d*)
but not by (*a*) except at very short chain lengths.

2.4 Elastomers

At this point we can introduce the concepts associated with
explaining rubber-like behaviour. The elastomers have a very low
elastic modulus and a very high elastic limit (they are easily
stretched, and can be greatly stretched without taking a permanent
set). They are long-chain polymers with occasional cross-links
between chains. To understand the explanation, originally put
forward by Marks, consider the photographs of models of a part
of hydrocarbon chain in various conformations. Fig. 9*a* (Plate)
shows the 'straight' conformation which occurs in crystalline
regions. Fig. 9*b* shows the same model arranged in a number of
random contortions. Clearly there is only one conformation, that of
Fig. 9*a*, in which the atoms M and N are the maximum distance
apart, but there are a very large number of conformations in which
the distance between M and N is much less than this maximum.
Thus in an amorphous polymer the most probable distance between
two atoms such as M and N will be much less than the maximum,
and if we took hold of M and N, pulled the chain out straight, and
then subjected it to random influences again, it would return to
a shorter-distance conformation.

Now suppose that atoms M and N are cross-linked by primary
bonds to similar atoms in another chain, and so on throughout the
material. Suppose further that the temperature is high enough to
allow chains to slide past each other without undue hindrances
from secondary bonds between them. On stretching, each chain is
straightened somewhat but the cross links do not allow complete
relative independence of whole chains. This straightening represents a slight approach to the orderliness of the crystalline polymer
(in fact a strongly stretched piece of rubber when examined by
X-ray diffraction does display a small degree of crystalline order
which disappears on releasing the extension), and the increase of
order is 'opposed' by the effect of thermal disturbance tending to
produce completely disordered arrangements. Since the unstretched rubber represented the natural state of disorder the rubber

will tend to return to its unstretched state; put differently, the individual lengths of polymer chain will tend to return to more probable 'less straight' conformations with, on the average, shorter rectilinear distances between given points on a chain.

2.5 The glass–rubber transition

It will be noted from the previous section that rubbery, or elastomeric, behaviour requires a temperature as high as that at which the polymer, but for the cross links, would be a viscous liquid or an easily deformed plastic solid. If the temperature is low, on the other hand, secondary forces between elements of neighbouring chains prevent their relative movement, and the elastic properties are like those of a hard organic substance, elastic deformation being due solely to changes in interatomic distances and spatial relationships. The change does not occur abruptly for reasons similar to those given in Section 2.3, but over the range of temperatures for which $\exp(-W/kT)$ changes by an order of magnitude. This change is called the *glass–rubber transition*, since below it the polymer is hard and amorphous, i.e. like a glass.

This term now tends to be used for the temperature around which any reasonably well-marked decrease of elastic modulus occurs, even though at a lower temperature the polymer is not very hard and at a higher one not very rubbery. It is not surprising, in view of the different kinds of secondary bond operating between chains, that many polymers show more than one such transition. These are usually referred to as α-, β-, γ- etc. transitions, α denoting the one which occurs at the highest temperature.

2.6 Significance of time scale

The process of deformation in polymers, consisting as we have seen in Section 2.3 of structural units slipping past one another or of secondary bonds being broken, by the chance concentration of an abnormal amount of energy, does not require all the units in a particular chain to move at the same time; since the chains are flexible the units can move one by one at different times. These movements are, of course, taking place whether the material is stressed or not; the higher the temperature the more frequent will be the interchanges of position, but as long as no stress exists the movements will be random and have no effect on macroscopic dimensions or stress.

If a solid is suddenly put in tension, the immediate (very short time) result is extension along the direction of the tension and a contraction in the plane at right angles; this is allowed by corresponding changes in interatomic distances, mainly between neighbouring atoms belonging to different chains. This is an elastic deformation: if the stress is promptly released the solid reverts to its original dimensions. For as long as the stress is maintained, however, the shuffling movements of the structural units will contain a slight preponderance of those which allow further deformation by relative movements of the molecules. If a constraint is maintained for a while and then removed the *elastic* deformation disappears, as before and the changes of position become entirely random, but (unless the molecules are cross-linked) the deformation corresponding to the relative movements by interchanges of position which have taken place is not reversed. There has only been a partial return towards the original dimensions: a *plastic* deformation has taken place. An obvious example of this is bitumen at room temperature which looks and feels like a glassy solid; a hammer striking it lightly will bounce off, but if the hammer is left lying on it for a few days it will sink in. Removing the hammer will then result in a very slight elastic recovery only. If the molecules are lightly cross-linked chains, however, the slow extension due to structural groups moving past one another will occur by straightening of chains rather than by complete slipping of one molecule past another. Removal of the stress after a time will allow an incomplete immediate recovery, the remainder will recover as the chains return to random conformations over a time comparable to the time for which the stress was applied.

It follows that if a lightly cross-linked polymer is at a temperature where its secondary inter-chain bonds will yield to a sustained strain but not to a brief one, then in the latter case it will behave as a hard solid, but in the former case as an elastomer. When it is strained for a long period and the constraint then removed, part of its recovery will take a time similar to the time over which the strain was applied. This is a glass–rubber transition in time rather than temperature, to this we shall return. An elastomer whose rubbery properties take several hours to develop at room temperature, say, if strained for a long period and then released, will apparently remain in the strained condition, but if warmed up to a temperature at which its response time is short, will appear to 'remember' its original shape and return to it rapidly. 'Heat-shrinking' and 'heat-set' are closely related; a material such as polyethylene terephthalate film, if stretched when hot and held so till it cools, will only shrink again noticeably if heated to a temperature approaching that at which it was 'set'.

Many of the time-dependent phenomena can be observed in ordinary rigid polyvinyl chloride on time scales varying from years to fractions of a second at temperatures between the freezing point and the boiling point of water.

An elastomer may have secondary cross links which will survive for a short time, say a millisecond, but not for a longer time, say a second. This is the behaviour of the polysiloxane 'bouncing putty' which can be moulded like plasticine but bounces when dropped.

In all the preceding discussion it must be understood that the actual time scales quoted are not unique; a 'long' time might be a few milliseconds, a 'short' time a few microseconds; at a lower temperature or with a different polymer 'long' and 'short' may be months and hours, respectively. If the time scale required to allow rubber-like behaviour is a matter of a few seconds, the rubber will be 'lazy', if only milliseconds, it will be 'snappy'.

At one extreme small long-term changes of dimensions take place in solids which appear stable by ordinary tests; at the other, a polymer which appears tough and resilient by ordinary tests, including impact tests, may shatter in a brittle manner under extremely high speed impact such as in compressed-air-operated equipment.

The usual simple theory of time-dependent mechanical behaviour (which is outlined below) gives a qualitative or illustrative picture, which can be empirically extended to fit actual behaviour in many cases. It is of interest here because it is parallel to the theory of dipole relaxation outlined in Section 4.1. It implies that the retarded extension resulting from a suddenly applied constant load, is proportional to the load multiplied by a decaying time-function, the same time-function whatever the load; similarly that the relaxing stress following a suddenly imposed constant strain can be similarly represented; though the retarded-strain time-function is not necessarily the same as the relaxing-stress time-function. This means that the retarded strain effects of any change of applied stress (negative as well as positive) can be added to those of any previous changes (superposition principle); similarly for relaxing stress if it is a change of strain that is imposed. This is true for small strains in many polymers and composites, but not all, some are non-linear even at low strains.

If we imagine a particular group of atoms bumping about among similar groups, the probability that an opportunity will occur (enough energy be acquired) to move past one of them depends on how long one waits. If a stress has just been established, rather more movements will take place adding to the strain, or relieving the stress as the case may be, than those producing the opposite result; the net increase of extension or relaxation of stress will be at a rate proportional to the rate at which groups find opportunity to move past each other. We assume (because it suits the intuitive model rather than from experience) that only a limited extension or relaxation can be accommodated by these movements, and that the rate at which they proceed falls in proportion to that which remains to take place.

We will now concentrate on strain at fixed stress (parallel arguments applying to stress relaxation at fixed strain), and will consider shear strain since it is a distortion without bulk compression.

Suppose the fixed stress q_s applied suddenly at time $t = 0$ produces an initial strain ϕ_i, which approaches a final steady strain ϕ_s; in line with what we have said we put

$$\frac{d\phi}{dt} = \frac{1}{\tau}(\phi_s - \phi)$$

giving

$$\phi_s - \phi = (\phi_s - \phi_i)\exp\left(-\frac{t}{\tau}\right). \quad . \quad . \quad . \quad . \quad 2.1$$

τ being a constant of proportionality called the *relaxation time*. The elastic compliance (strain/stress) J will have an initial value $J_i = \phi_i/q_s$ and an ultimate long-time value $J_s = \phi_s/q_s$ and it is convenient for future reference to write eqn. 2.1 as

$$\phi = q_s\left\{J_i - (J_s - J_i)\exp\left(-\frac{t}{\tau}\right)\right\} . \quad . \quad . \quad . \quad 2.2$$

It will be obvious that relaxation times are very dependent on temperature, since the probability of changes of position depends on acquiring a given amount of energy; τ is therefore likely to be proportional to something like $\exp(W/kT)$ where W is the energy needed to activate the particular type of inter-chain movement we are considering.*

* This relationship with temperature appears to be too simple, however. A widely used one (Ferry, 1970, and Billmeyer, 1970, Chapter 6c) is equivalent to

$$\tau = A \exp\{b/(T - T')\}$$

where b is about 2070 and T' is a temperature some 51·6° lower than the value of T_g defined by the change in thermal expansion coefficient which marks the temperature at which molecules begin to occupy more than a minimal volume. A depends on the polymer and is equal to

$$10^{-17\cdot4} \text{ (value of } \tau \text{ at } T_g)$$

B

We have now moved from the concept of Sections 2.3–2.5 that increase of temperature increases the number of units which are 'free', to the concept that it increases the frequency with which any unit has an opportunity to move, and therefore decreases the relaxation time τ. Because of the exponential relationship of τ to temperature we are much more aware of the effects of temperature than of the effects of time-scale; a few degrees change of temperature may be equivalent to extending the period of observation by ten times.

Illustrating the ideas we have discussed in this and the preceding sections, a plot of elastic compliance against temperature, passing through a glass–rubber transition, will look like one of the curves, say (ii), of Fig. 10. If the measurements are repeated but taking ten times as long to measure the extension at each stress, the curve will be found displaced to low temperatures as curve (i). Conversely, if the measurement process is greatly speeded up, a curve such as (iii) will be obtained.

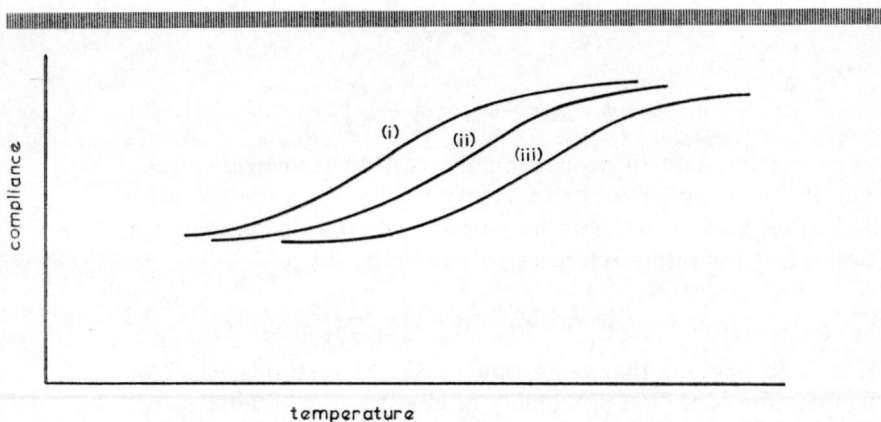

Fig. 10 Typical variation of elastic compliance with temperature in the neighbourhood of the glass–rubber transition
(i) Measurements over long periods
(ii) and (iii) Measurements over shorter and shorter time-scales

A convenient method of investigating and displaying this is by applying sinusoidal stress. The component of strain which is in phase with the force then measures the real part of the 'dynamic' compliance, and the frequency of oscillation measures the time scale. Curves (i), (ii) and (iii) of Fig. 10 would then correspond with three frequencies increasing in that order. There is also a component of strain in phase-quadrature with the stress; the compliance is now a dynamic compliance and must be represented as a complex quantity.

There is a very general relationship between the response of a linear relaxing system to a stepwise change, described by eq. 2.2, and its response to a sinusoidal oscillation. The response to an applied sinusoidal stress

$$q = q_0 \exp(j\omega t)$$

can be represented by

$$\phi = Jq$$

where ϕ and J are complex and J is given by

$$J = J' - jJ'' \quad . \quad . \quad . \quad . \quad . \quad . \quad . \quad 2.3$$
$$J' = J_i + (J_s - J_i)/(1 + \omega^2\tau^2) \quad . \quad . \quad . \quad . \quad 2.4$$
$$J'' = (J_s - J_i)\omega\tau/(1 + \omega^2\tau^2) \quad . \quad . \quad . \quad . \quad 2.5$$

The justification for this is well known (see, for example, Gross, 1948 or Daniel, 1967), a justification by analogy with circuit equations is given in the Appendix.

The model we have adopted (eqn. 2.2) displays a creep, taking place after the application of a steady load, which approaches an asymptotic value as illustrated by the full line in Fig. 11. It also displays a response to sinusoidal stress (or strain) represented by a complex compliance. The real and imaginary parts of this, J'

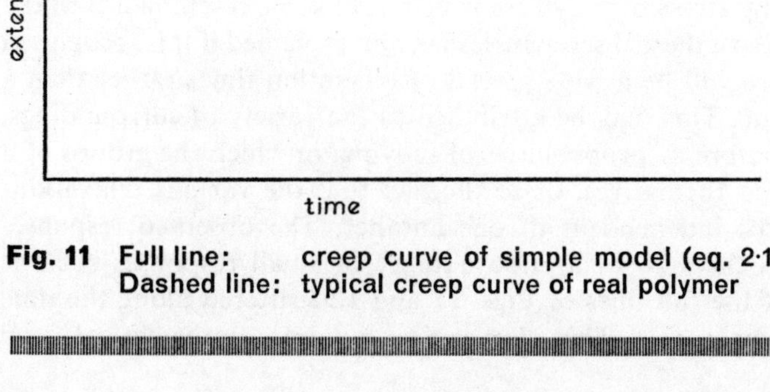

Fig. 11 Full line: creep curve of simple model (eq. 2·1 or 2·2)
                Dashed line: typical creep curve of real polymer

and J'', are described by equations which are similar to those for the components of permittivity ε' and ε'' of Section 4.1; the latter are shown in Fig. 15 as functions of log(angular frequency), the curves for J' and J'' are of the same shape. It is usually more convenient in the case of J' to plot its logarithm and this is done in Fig. 12 (full line). J'' is a measure of the energy loss per cycle.

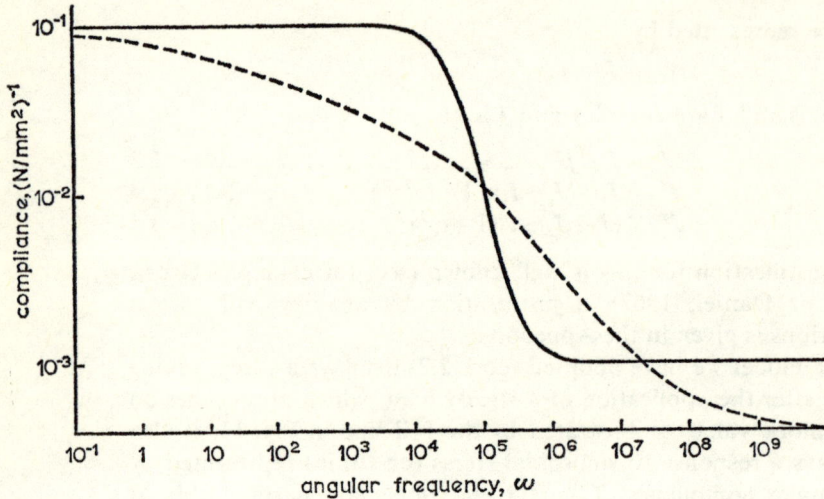

Fig. 12 Full line: real part of complex compliance from simple model (eq. 2.4)
          Dashed line: real part of complex compliance of a vulcanised rubber

Real materials are often qualitatively similiar to this model but not quantitatively so. The dashed line in Fig. 11 shows a typical creep curve, which has no asymptote; the dashed line in Fig. 12 which is typical of a vulcanised rubber indicates the response to oscillating stress of polymers which have some resemblance to the model. Both these discrepancies may be explained if it is recognised that there will be a wide spread of relaxation times rather than a single one. This may be attributed to the variety of surroundings, and therefore of probabilities of moving, in which the groups of a chain find themselves, or to the fact that the various relaxations cannot be independent of one another. The observed responses are thus the sum of a large number of small responses such as those of the full lines of Figs. 11 and 12 scattered along the time or frequency axis. This idea can be put into mathematical form

(Gross, 1948), but dong so adds little to understanding it. It should be pointed out that this explanation involves the assumption that there is some structural feature of a polymer that has a relaxation time as short or as long, from nanoseconds to years, as the duration of any observation that may be made. The question of distributed relaxation times is taken up again in Section 4.1.

The general shape of the curves of Fig. 12 is like a mirror image of those of Fig. 10, because eqns. 2.4 and 2.5 are symmetrical in ω and τ, and since τ varies approximately exponentially with temperature a change of T is roughly like a proportionate change in log ω, apart from the effect that temperature may have on $(J_s - J_i)$. From eqns. 2.4 and 2.5 and Fig. 15 it is clear that J'' goes through a maximum at the same frequency and approximately the same temperature as J' has its greatest rate of change, this is still the case with a symmetrical spread of relaxation times. A glass–rubber transition can therefore be identified by a maximum in the loss J'' accompanying mechanical vibration, which is sometimes more convenient to observe than a rapid change of compliance.

For predicting creep over long periods it is convenient to plot both extension and time logarithmically, since this plot does not usually have marked curvatures. (This has led to the rough empirical equation

$$\phi \propto t^{-n}$$

over limited ranges of t, n being a small quantity such as 0·15). For some purposes it is useful to replot such results as a family of stress–strain curves each referring to a particular duration of the stress. Many examples of creep, relaxation and isochronous stress–strain curves of a number of common thermoplastics are given by Ogorkiewicz (1970*a*). The creep properties of reinforced plastics are discussed elsewhere by the same author (1970*b*). Obviously creep is very temperature-dependent and full regard must be paid to this.

The preceding discussion is not meant to be exhaustive, and does not suggest that all aspects of polymer behaviour have been explained; it does imply that such phenomena as plastic flow, creep and memory can result from quite normal chemical structures. It has been tacitly assumed that the polymers we are concerned with are mainly amorphous, and the stresses and strains uniform. With largely crystalline polymers the situation is more complex because crystalline and amorphous regions have different properties, and their relative proportions may be altered by straining. Crystalline polymers may display strain-hardening and the property of cold drawing (see introductory part of Chapter 9).

2.7 Mechanical strength

The properties in the closely related group which includes tensile strength, yield stress, cross-breaking strength, flexural fatigue resistance, impact strength etc. have not been related in detail with chemical and physical structure, basically because the microscopic processes associated with rupture are the least understood of all.

Other things being equal, longer chain lengths produce higher tensile strengths, cross-breaking strengths, yield stresses and flexural fatigue. Longer chain length generally means higher fabrication temperatures, with greater possibility of decomposition, and some compromise with mechanical properties may be necessary. Increased crystallinity naturally increases density, and may increase elastic modulus and yield stresses by factors of between two and four, with a decrease in elongation at break and often in flex resistance. It is a rough but by no means reliable rule that stronger secondary inter-chain bonds, and cross-links, increase elastic moduli, yield stresses and tensile and cross-breaking strengths. Impact strength does not parallel the others because increased tensile strength and increased rigidity tend to work in opposite directions. Cross-linking increases stiffness, toughness and mechanical strength but very highly cross-linked materials may be undesirably hard, brittle and poor in impact strength. Cross-linking also reduces crystallinity, which may tend to cancel the benefits. It seems possible that in the future the advantages of cross-linking and of crystallinity could be combined; various crystalline 'ladder' polymers whose main chains are straight (the stiles of the ladder) with regularly spaced cross-links (rungs) have been produced; none of them are yet in practical use.

Time effects are important in relation to ultimate failure as well as in relation to the properties discussed in Section 2.6. For example, if creep is observed at a stress which is, say, one-third of the short time breaking stress, the rate of extension may not continue to decrease as in Fig. 11 but pass through a point of inflection and thereafter accelerate until rupture occurs (Fig. 13). Creep-rupture curves, i.e. plots of breaking stress against logarithms of time to break, are usually fairly straight lines, the breaking stress dropping by a factor of two, three or more over a period of a year (Ogorkiewicz, 1970a).

The qualities needed to avoid failure in a particular application may need careful analysis, or trial and error. It is not uncommon for an engineer dealing with a situation where distortion is inevitable to think he wants a 'stronger' material when in fact he wants a

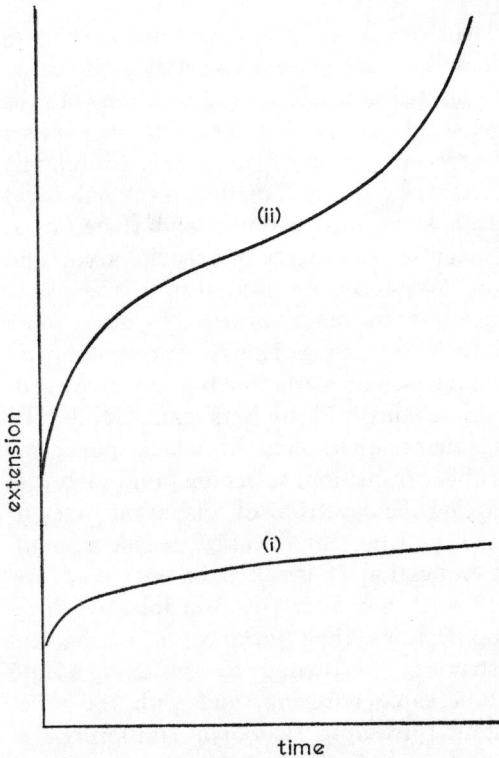

Fig. 13 Creep at low and high stresses

(i) At low stress creep takes place at a continually diminishing rate, no failure resulting

(ii) At high stress the rate at first diminishes then increases, ultimately leading to failure

more flexible one, which may actually have a lower tensile strength than the one which fails. Even the mechanical loss factor may be important in some circumstances.

There are a number of mechanical qualities, more easily appreciated than explained, for example toughness and cut-through, which are presumably combinations of the properties we have discussed and which may be vital to the success of some applications but can only be evaluated empirically.

There is interest in using insulating materials at temperatures as low as 4 K (liquid helium), at which polymeric materials are often so brittle as to be useless. There is little systematic information about this; polytetrafluorethylene and polyimides appear to be usable, and composites such as epoxy-glass laminates have been employed.

2.8 Effects of plasticisers

The importance of plasticisers is very great and must be empha-sised here though this book is not the place for a detailed discussion of their technology. The plasticiser forms a solid solution with the polymer, and almost any substance which will do this can act as a plasticiser, though not necessarily a satisfactory one. Commonly used ones are mostly esters: the esters and diesters of phthalic acid with octyl, decyl etc. alcohols, various aromatic phosphate esters, butyl oleate and butyl stearate, and esters of sebacic, azelic and adipic acids. Many other substances are used also.

Their principal purpose is to make brittle polymers more flexible and/or facilitate their processing. This they presumably do by reducing the aggregate inter-chain attraction (loosely by provid-ing 'lubrication' between chains). Plasticisers can usefully be considered as lowering the temperatures at which particular phenomena, e.g. glass–rubber transition, softening point or brittle point, appear; or as moving the spectrum of relaxation times to lower temperatures. This is an over-simplification, tensile strength is usually reduced and elongation at break point increased, for example. Their presence does not alter the principles we have discussed, but the modifications they produce in mechanical properties may be very striking; as is obvious by comparing a rigid polyvinyl chloride with a cable covering, and with the sheet materials used for domestic furnishing. (Copolymerisation is also used as a means of increasing flexibility.) Plasticisers, if not properly chosen, may evaporate during service, resulting in embrittlement and shrinkage. The usual plasticisers, being dipolar, tend to increase dielectric losses, and, whether dipolar or not, to shorten dielectric relaxation times of dipolar molecules already present.

Crystalline polymers are more resistant to solvents than their amorphous counterparts, and hence are less affected by plastic-isers.

Chapter 3

Chemical stability

3.1 Nature of thermal limitations and methods of estimating allowable operating temperature

The operating temperatures of the inorganic insulants—porcelains and glasses—are limited by softening, or by measurable and reversible changes of conduction, dielectric loss or dielectric strength, or by the danger of fracture due to differential thermal stresses occasioned by uneven heating, changes of section or constraint by other materials. These are limitations which can be determined by operating tests and overcome to some extent by choice of materials or by careful design.

Organic materials on the other hand suffer well-known types of irreversible changes at high temperatures; generally the temperature limit which will restrict deterioration to what is acceptable over the normal lifetime of equipment is lower than that imposed by immediate changes such as softening, except when the temperature rise in service is of short duration. Typical obvious symptoms are shrinkage, hardening, spontaneous cracking or crazing, loss of strength, embrittlement, discolouration, distortion and, in extreme cases, charring These effects are generally due to, or accompanied by: loss of weight resulting from evaporation of volatile components; oxidation or pyrolysis* to form gases or volatile substances such as CO, CO_2, water and organic compounds of low molecular weight; cross-linking (over and above any intentional cross-linking) due to continuation, of the reaction by which the polymer was first formed, or to condensation reactions† between the products of oxidation, or to the formation of oxygen

---

* Pyrolysis is the splitting up of a compound into smaller molecules without involving any other substance, such as oxygen, from the air
† Condensation reactions are explained in Section 2.1

B*

links, or to more complex sequences of reactions Hardening and embrittlement imply that the elastic modulus has increased, and generally this seems to occur by movement of the glass–rubber transition(s) to higher temperatures, rather than by a general increase of modulus at all temperatures.

It is natural and probably correct to suppose that much deterioration results from oxidation by air, but ageing experiments in gases such as hydrogen or nitrogen do not always give startlingly different results from those in air, suggesting that pyrolytic reactions go on at temperatures little higher than oxidation reactions; or that small quantities of air or water are sufficient to produce ageing.

To compare the suitabilities of materials for use at high temperatures it is desirable to have some quantitative measure of deterioration which can be employed in the laboratory. The simplest change to measure is the rate of loss of weight at a given temperature; comparisons are often made on this basis, but weight is not one of the attributes of direct concern, nor is loss of it a reliable measure of the amount of degradation that has occurred, since loss by evaporation or oxidation to water or oxides of carbon may be partly offset by gain of weight due to formation of involatile oxidation products Nevertheless, the method is simple, convenient and often used. An elaboration of it, now becoming known as thermogravimetric analysis, consists basically of raising the temperature of the sample at a defined rate and plotting weight loss against time (usually on a scale of hundreds of hours) and attributing appropriate explanations to any non-uniformity or sudden curvature of the plot (Brown *et al* , 1973; Toop, 1971).

Loss of tensile strength is frequently used for comparing cellulosic fibres and fabrics, but it is not so convenient for more massive forms of material, and is not usually appropriate for materials used for coating, impregnating and sealing.

Chemical assay of acids or other oxidation products is useful for hydrocarbon oils but not very practicable for solid insulation.

In general the most valuable results are those obtained by simulating (however crudely) service conditions, using a knowledge of the type of failure which occurs in the kind of service concerned to devise a suitable criterion of endurance to these conditions. Examples are the twisted wire and motorette tests (Sections 6.2.5.4 and 6.2.5.5.).

In tests designed to assess the useful life of insulating materials we usually want to predict the highest temperatures at which they will give satisfactory service for a period of years; which may correspond to 1000, 10 000 or 100 000 h according to the type of duty cycle the equipment is intended for. Test periods of 10 000 h or more involve much too great a delay in reaching design decisions, and to shorten this a higher test temperature must be used. It is found, however, that the order of merit obtained from short time high temperature tests is not always the same as that found from longer time, lower temperature tests. One obvious reason for this would be the differing degree to which the reactions causing degradation in different materials are accelerated by temperature; a test method is therefore needed which takes account of this. This is based on Arrhenius's Law which can be broadly understood by an argument similar to the one we have already used in connection with the freeing of secondary bonds: a definite amount of energy, the *activation energy*, is necessary to initiate the splitting up of a molecule or a reaction, such as oxidation, between molecules. The number of molecules which acquire this energy, W_e say, in one second is approximately proportional to $\exp(-W_e/kT)$* so that the rate of the reaction should vary with temperature as

$$\text{rate} = A\exp(-W_e/kT)$$

if there is an ample supply of the reactants and rapid removal of the products.

We now make a number of sweeping assumptions, notably the following: that the life of a piece of insulation from the start of the test to the moment when the specimen is judged to have failed (by having lost a given amount of weight, broken down mechanically or electrically, or whatever criterion is applied) is the result of the unhindered progress of a single reaction to a particular stage. It would then follow that

$$\log(\text{life}) = A + B/T \qquad . \quad . \quad . \quad . \quad . \quad 3.1$$

A and B being constants which could be determined from observations at a number of temperatures. This equation provides a means of extrapolation to lower temperatures and longer times than the test periods. Even though the assumptions are not strictly true the approximately exponential relationship is still likely to

---

* This is written in terms of the energy required to break a single bond, or cause one pair of atoms to react. In chemical contexts it is more usual to write the energy W_M needed to cause one mole of material to react, and the expression becomes $\exp(-W_M/RT)$ where R is the gas constant. This amounts simply to multiplying top and bottom of the fraction by the number of molecules in a mole

hold. If several reactions are involved in sequence it is likely that only one of them controls the rate. If some physical process like diffusion of oxygen into the insulant determines the rate of reaction this also is likely to obey an equation of similar form. In any case no better method of extrapolation has so far been found.

Generally three or four ageing temperatures are chosen such that the lowest is likely to produce failure in about 5000h, and the highest in not less than 100h; ovens of constant and uniform temperature are required. Several samples must be aged at each temperature to avoid reaching misleading conclusions because of the scatter of results between one sample and another. They are removed from the ovens at regular intervals and subjected to any desired treatment such as a humidity chamber, and to any chosen test such as an electric strength proof test.

If eqn. 3.1 is justified, the logarithms of the times to failure (counting, of course, only the time spent at the ageing temperature), plotted against the inverse of the thermodynamic temperatures (°C plus 273) will lie on a straight line. Because of the scatter the results will look like Fig. 14, and a statistical procedure is normally used to ensure that the 'best' line is drawn through the points (see Section 6.2.5.3 and Nelson, 1971/72).

If the results of two materials are like the two lines (i) and (ii) in Fig. 14 there can be little doubt of the conclusion, but if they cross, as do lines (iii) and (iv), or if the observations will not fit a straight line, it is very desirable to do check tests before accepting the order of preference.

In practice it is unwise to attempt to draw firm conclusions about the relative service lives of two materials unless the results of this type of test differ by at least a factor of 2:1 in life for a given temperature (corresponding to several deg C for a given life). Unfortunately, comparisons between different laboratories carrying out nominally the same test on the same materials have shown that this discrepancy can occur between careful experimenters (Mahon, 1969; Appleby *et al.*, 1970, Goldenberg, 1962). Test carried out by one person at the same time with the same equipment may be more consistent.

The elaboration of these tests has been developed because of the variety of solid insulants available and the wide range of temperatures concerned, particularly in rotating machines. The variety of possible combinations of materials, such as wire enamels and impregnating varnishes in motor windings, involves many possibilities of the thermal endurance of one material being adversely affected by the proximity of another.

Liquid insulants on the other hand, are so few that the allowable temperatures in combination with cellulosic materials and in various circumstances are well established by experience.

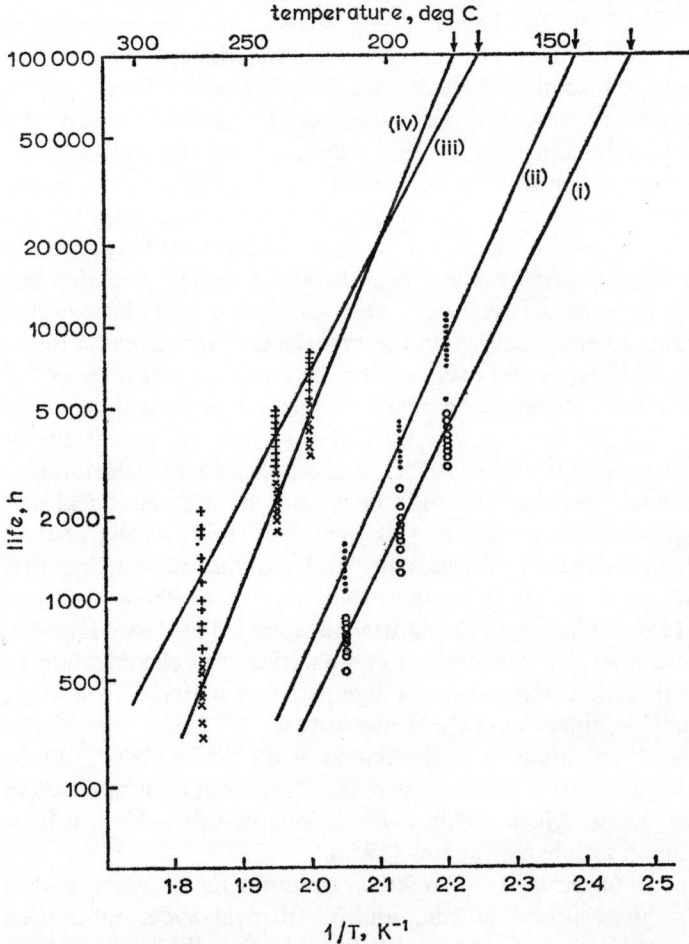

Fig. 14 **Typical results of ageing tests on four insulating systems**

Time to failure is plotted logarithmically against reciprocal thermodynamic temperature, with corresponding temperatures in deg C on upper scale. Temperatures at which expected life is 100000 h are shown by arrows. (Observations do not always fit a straight line as accurately as these. It is common to reverse the direction of the 1/T scale so that temperature increases from left to right.)

3.2   Thermal classification system

This system of classification relates primarily to materials for use in rotating machines and transformers. At the time when available materials were limited to natural fibres, drying oils and resins, natural rubber, phenol formaldehyde resins, mica, porcelains and glasses they fell fairly clearly into the four classes O, A, B and C of the list below. These letters were not regarded as synonymous with the temperature limits mentioned, but as labelling types of insulation in order of increasing thermal stability, the temperatures being approximate guides. The operating duty of a piece of equipment often implies a cycle of temperature variation in which high temperatures occur for only limited periods; in any case temperatures in machines are often difficult to measure and the designer may not have the means to know what they are. The choice between classes A and B in transformers and machines was thus directly associated with types of design and duty cycles, depending to some degree on the experience, caution and circumstances of designer and user rather than on direct measurement of temperatures. Moreover, a small proportion of class A material (such as paper backing for mica flakes) may be acceptable at temperatures well above 105°C, while a degree of deterioration which would not be allowable in regions of high electrical and mechanical stress may be harmless in regions which involve neither.

The appearance of silicone materials led to the introduction, first in USA, of class H; later, the availability of synthetic resins of better heat stability than the natural oil resins led to classes E and F.

Where materials are used in combination, the classification of the combination may not be as high as either material separately, or it may be higher than the worse of two.

A very brief summary of the classes of BS 2757 or IEC Publication 85 and of the materials listed as examples in their appendices is given below. More detailed discussions of this subject will be found in an article by Snadow (1952).

Class Y (formerly O) up to 90°C: Unimpregnated paper, cotton or silk, urea formaldehyde, aniline formaldehyde, vulcanised natural rubber, and various thermoplastics limited by their softening points.

Class A up to 105°C: Paper, cotton or silk impregnated with oil or varnish, or laminated with natural drying oils and resins or phenol formaldehyde; esterified cellulose fibres, polyamides and a variety of organic varnishes and enamels used for wire coating and bonding.

Class E up to 120°C: Phenol formaldehyde and melamine formaldehyde mouldings and laminates with cellulosic materials; polyvinyl formal, polyurethane, epoxy resins and varnishes; cellulose triacetate, polyethylene terephthalate, oil modified alkyd.

Class B up to 130°C: Inorganic fibrous and flexible materials (mica, glass or asbestos fibres and fabrics etc.), bonded and impregnated with suitable organic resins; shellac bitumen, alkyd, epoxy, phenol- or melamine-formaldehyde.

Class F up to 155°C: As class B but with resins approved for this class: alkyd, epoxy, silicone-alkyd and certain others.

Class H up to 180°C: As class B but with silicone resins; silicone rubber. (Note that a number of high temperature resins have appeared since this classification was drawn up.)

Class C: Mica, asbestos, ceramics, glass, alone or with certain inorganic binders or silicone resins; polytetrafluorethylene.

The original four-class system was natural and useful at the time when it was widely believed that no organic resin could withstand temperatures higher than 130°C even in combination with inorganic materials. Now it has become clear that there are a number of ways of increasing stability in purely organic materials, for instance cross-linking, substitution of fluorine for hydrogen, the presence of aromatic groups, and strong hydrogen bonding. Thus the more highly cross-linked alkyds are more stable to oxidation than the less cross-linked ones, polytetrafluorethylene is much more stable than polyethylene, aromatic polyamides than alkyl polyamides, and polyimides than polyamides. The situation became more complex in a way that the addition of three more classes did not permanently resolve, while it is becoming more more important than ever to make the optimum use of materials for compactness and economy.

In the USA the Institute of Electrical & Electronics Engineers' publication No. 1 (1969) recommends assigning a *temperature index* to each material, based on its life in respect of some property in particular environmental conditions (a material may have different indices in respect of different properties or conditions). These indices are to be determined by experience or by test procedures in comparison with materials of established indices, and each is a number preferably chosen from the series 90, 105, 130, 155, 180, 200. Although these numbers, apart from the last, are reminiscent of the international classification they do not represent limiting temperatures but a scale for comparing one material with another.

The 1958 seven-class international system must be considered out of date but it has proved very difficult to agree on a new one, some holding that a much more elaborate system is needed, taking into account other factors than temperature, others that such elaboration would be ineffective and that efforts should be confined to establishing uniform methods of evaluation. A system making use of several factors is, in fact, being prepared. It must be recognised that the variety of available materials and their combinations, mixtures, copolymers, reinforcements, methods of curing and so forth, presents formidable problems of definition and identification, as well as classification, so that the user who is not willing to accept the machine manufacturer's judgement must enter into discussion with him on an expert plane, and on this plane the existence of a classification may be irrelevant.

3.2.1   Note on thermal conductivity

In principle this is a factor to be considered in design, since the greater the thermal conductivity of a piece of insulation the greater the quantity of heat that it will conduct away without exceeding the allowable temperature, and therefore the higher the rating of the equipment concerned. In practice, however, it is a factor that is rarely taken into account in choosing insulating materials; only in very simple thermal systems such as cables and in some types of electronic equipment, where heat removal is very critical, is a serious effort made to predict and observe its effects.

The measurement of thermal conductivities, particularly of non-metals, is far more difficult than the measurement of electrical conductivities. In the first place there is no method of measuring directly the local temperature of a material such as a plastic or ceramic when a temperature gradient exists, because (*a*) the introduction of a thermojunction disturbs the temperature distribution from which the conductivity is to be calculated, and (*b*) the thermal resistance between the material and the junction is likely to be comparable with, or greater than that between the couple and the surroundings, so that its readings are worthless. These difficulties are usually avoided by measuring the temperatures of metal blocks in supposedly intimate contact with the material under test. However, if the material is thin it is likely that the thermal resistance of the small local gaps between metal and specimen will be significant in comparison with that of the specimen itself. If the specimen is thick, yet of reasonable dimensions, the correction for heat loss to the surroundings, which is difficult to determine anyway, becomes larger than the quantity to be measured. Indeed, many figures quoted for thermal conductivity appear improbable;

reliance should only be placed on those provided by establishments highly skilled in this type of measurement and prepared to go to considerable trouble in methods of experimental verification.

The uncertainties involved in measurement of heat transfer by conduction apply equally to predicting temperatures when the conductivities are known. The thermal path between a conductor in a slot of a machine and the walls of the slot, for example, involves at least one air gap, possibly many. (The thermal resistance of a thin air gap is roughly the same as that of a plastic sheet ten times as thick.)

The range of variation between different organic solids is not great, while the temperature drop across the insulator is usually much less than the total drop to ambient temperature. Consequently, the maximum temperatures which different materials will stand assume more importance than the effects that thermal conductivity differences have on the temperatures produced. There may be a case for giving this question more attention, but a more detailed and precise approach would be required than has been practicable hitherto.

In oil-cooled equipment, such as power transformers, heat transfer is by convection. Natural convection coefficients vary with the properties of the liquid—thermal conductivity k, expansion coefficient a, specific heat per unit volume c and kinematic viscosity v—approximately according to the function

$$\left(\frac{k^3ac}{v}\right)^n$$

(where n is $\frac{1}{3}$ or $\frac{1}{4}$; see, for example, Stoever, 1941). Thermal conductivities do not vary more than 10–20% between different transformer oils or chlorinated diphenyls (Clark, 1962) while viscosities can vary enormously, factors of 10:1 between different fluids at the same temperature and of 100:1 for a particular fluid over its working temperature range are common. Again thermal conductivity is not a significant factor in the choice of material to be used, except perhaps in the case of impregnated paper cables and capacitors where convection does not occur.

For these reasons, thermal conductivity of organic materials is not mentioned in Chapters 7–10. In certain types of electronic equipment, on the other hand, removal of heat with strictly limited temperature rise is crucial to operation. These usually involve ceramic insulators, and the differences between one ceramic and another, or between different forms or densities of the same ceramic may be highly important.

For solid plastic materials thermal conductivity generally lies between 0·15 and 0·25 w/mK; for mineral-filled and glass-reinforced materials the value may be two or three times as high; for imperfectly impregnated fibrous materials it may be considerably lower.

3.3 Effects of solvents

High molecular weight polymers do not generally dissolve completely except at temperatures near to their melting points; but many take up solvent and swell on immersion or on exposure to vapour. Insulation which may be exposed to traces of solvent vapour should be chosen with care; the quantity of solvent may be quite inadequate to cause swelling but, over a period of time, may cause crazing, and possibly cracking of any portion of the polymer under tensile stress ('environmental stress cracking'). Solvent effects are reduced by crystallinity (e.g. isotactic polypropylene is much less affected than low density polythene), and by cross-linking. Highly cross-linked materials, such as phenol formaldehyde resins, can be used under oil.

3.4 Stability under irradiation

Any radiation more energetic than the visible spectrum can cause important chemical changes and affect, usually adversely, organic insulants. The effects of sunlight are in many ways like the effects of heat. Oxidation, shrinkage, cracking etc. of the hydrocarbon polymers can be countered by opaque fillers, usually at the expense of some other property. Some polymers, such as polyvinyl chloride, are little affected.

The much more energetic radiation associated with nuclear reactions—high energy electrons, protons, neutrons and γ-rays— all have fairly similar effects, the immediate cause of the effects probably being the same in every case, namely, the energetic secondary electrons produced by the primary radiation. In the absence of air the principal effects are a combination of chain-breaking (scission) and cross-linking. In a number of hydrocarbon polymers including polyethylene, polypropylene and polystyrene, the cross-linking is predominant, at least for moderate doses, usefully so in the case of polyethylene which can be cross-linked sufficiently to make it a soft rubber above its normal melting point, and give it some degree of stability of shape up to 200°C (Black and Charlesby, 1960).

There is a small but growing requirement for electrical equipment to operate within the radiation screens of nuclear reactors and particle accelerators, where the flux of neutrons is sufficient to degrade many organic materials and would shorten the life of the equipment. Waddington (1962) has assessed the probable acceptable radiation dose for a representative variety of polymers, filled and unfilled, from various sources of data and has compiled lists covering moulding materials, rubbers, varnishes, liquids and electronegative gases. A review of radiation effects on thermoplastic polymers, oils and cellulose has been made by Black and Reynolds (1964).

Chapter 4

General features
of electrical behaviour

It is natural to expect that the important electrical properties of an insulating material will be characteristic of that material or of its major constituents (as the important mechanical properties usually are), but this is true only of permittivity and high frequency dielectric loss. Electric strength, d.c. conductivity and low frequency dielectric loss are highly sensitive to small quantities of moisture and other impurities; moreover, the observed value of electric strength is greatly dependent on the manner in which it is tested.

We will deal with permittivity and dielectric loss first.

4.1 Permittivity and dielectric loss

Every textbook of electricity explains that the difference between the permittivity of a dielectric and that of free space is due to the restricted movements of charges within the dielectric, positive charges moving with the electric field and an equal quantity of negative charges moving oppositely, so that there is no net charge anywhere within the dielectric, but there is a net positive charge at the surfaces where the positive direction of the field emerges and a negative charge at surfaces where the field enters. Thus the field within the dielectric is produced by a field outside it which is larger, the normal components having the ratio given by the relative permittivity. This is called polarisation. The charges are bound in the dielectric, they cannot move throughout it or they would produce conduction, not polarisation. This polarisation is generally of three basic kinds:

(*a*) Relative movements of electrons and nuclei of atoms, so that the nuclei are no longer at the centres of the electron 'orbits'. This occurs in all materials and when there is no other polarisation it produces a relative permittivity of about two.

(*b*) In an ionic or partly ionic crystal, or a glass, relative movement of positive and negative ions: this together with (*a*) generally produces relative permittivities of five or more (very much more in the case of certain ceramics).

(*c*) In a molecular substance containing dipolar groups (see Section 2.2) turning of the dipoles slightly in the direction of the field, so that the positive 'ends' of the dipoles point rather more with the field than against it: this produces relative permittivities ranging from two if the dipoles are sparse to values such as 36 for nitrobenzene and 80 for water.

The permittivities and losses in ceramics and glasses are attributed to ionic polarisation because the partly ionic character of the bonds between atoms must produce a shift of charge distribution accompanying any relative movements; an applied electric field will superpose on the thermal vibrations of the partial ions a slight average displacement of positives in one direction and negatives in the other. Losses may be attributed to energy dissipated in various ways by moving ions from one position to another and back again (see Section 4.2.3.2).

The rotation of permanently dipolar groups provides well-founded explanations of the behaviour of many organic materials in practical use. In visualising this process we do not think of an array of otherwise stationary dipoles turning with field when it is applied, but dipoles pointing in all directions and continually jumping from one orientation to another by thermal agitation. The polarisation which appears when a field is applied is a relatively small average favouring of the direction of the field. There is not necessarily a restoring 'force'; the tendency to revert to random orientation opposes the tendency of the field to align the dipoles and allows a polarisation varying in proportion to the applied field. The dipolar polarisation produced by a field as high as $10 \, \text{kV}/\text{mm}$, nearly the highest used in engineering practice, would be less than one hundredth of that caused if all the dipoles were aligned parallel.

The polarisation movements take time; type (*a*), and many cases type (*b*), take place in time shorter than the period of oscillation of even millimetre wavelength radiation, and so for our purpose are instantaneous. Type (*c*), dipole rotation, is very similar to the molecular movements giving rise to mechanical relaxation discussed in Section 2.6 and is subject to a similar relaxation time. To understand how the time needed to establish dipole polarisation arises perhaps the easiest model to visualise is something like a spring-loaded rotary selector switch with a number of stable orientations; it is capable of being moved from one orientation to another by provision of enough energy to overcome the potential energy barrier provided by the spring. The barriers in the case of the dipole may be hindrances from neighbouring parts of the molecule, electrostatic interactions or other forms of restraint. As in Section 2.6 we assume an energy W_0 has to be imparted to overcome the barrier between one orientation and the next, so that the probability of jumping is approximately proportional to the familiar expression $\exp(-W_0/kT)$. The average preference in favour of the direction of the field can only take effect when enough jumps have occurred.

If we suddenly apply a field E the dipolar polarisation will grow to an equilibrium value at which the rate of polarisation produced by the field is equal to the rate at which it is destroyed by the thermal agitation. Suppose the equilibrium polarisation is a fraction m, say, of the polarisation which would result if all the dipoles were aligned parallel to the field, the detailed theory (Smyth, 1955; Bottcher, 1952; Wyllie 1960) shows that

$$m = \mu E_d/3kT$$

μ being the dipole moment (of the order of 10^{-29} Cm) and E_d the average field acting on the dipoles themselves which is of the same order as, but not equal to, the field E measured in the ordinary way. The rate at which the field creates polarisation from the nearly randomly oriented dipoles (nearly random because m is small) is practically independent of the polarisation already produced, while the rate at which polarisation is destroyed by random motion must be proportional to the polarisation existing. If we write the constant of proportionality as $1/\tau$ (anticipating that τ will become the relaxation time of the dipole polarisation) we get

$$dP/dt = (\text{constant}) - P/\tau$$

P being the instantaneous value of polarisation. Solving this with the condition that the final value of P is P_S gives

$$P = P_S\{1 - \exp(-t/\tau)\} \quad . \quad . \quad . \quad . \quad . \quad 4.1$$

This equation refers only to one dipole component of polarisation, the electronic component and any other dipole components are to be added to it.

A fuller discussion of the argument sketched above is given by Frohlich (1949). We have omitted a great many things which are necessary to understand the relationship between dipole moment, polarisation and field strength, or the interaction between different contributions to polarisation; these have been covered in various reviews (Cole, 1961; Daniel, 1967) as well as in the texts already mentioned. The relationship between polarisation and permittivity is explained in any textbook of electricity; here we will just state that if a potential v_s is suddenly applied to electrodes attached to the faces of a simple dipolar dielectric a charge Q_i due to electronic polarisation will appear in a negligibly short time, and this will grow to a final value Q_s as the dipole polarisation develops, which may be a matter of hours, seconds, microseconds or nanoseconds. From eqn. 4.1 it follows that the charge Q at any time after applying V_s will be given by

$$Q_s - Q = (Q_s - Q_i) \exp(-t/\tau) \quad . \quad . \quad . \quad . \quad 4.2a$$

and the current flowing to the electrodes will be

$$I = (Q_s - Q_i)(1/\tau) \exp(-t/\tau) \quad . \quad . \quad . \quad . \quad 4.2b$$

The permittivity will have an initial value $\varepsilon_i = Q_i/V_s$ and an ultimate value $\varepsilon_s = Q_s/V_s$; we now rewrite 4.2a as

$$Q = V_s\{\varepsilon_s - (\varepsilon_s - \varepsilon_i) \exp(-t/\tau)\} \quad . \quad . \quad . \quad . \quad 4.3$$

This is identical in form with eqn. 2.2 and the response to a sinusoidal voltage

$$V = V_0 \exp(j\omega t)$$

follows similarly (see also Appendix). We write $Q = \varepsilon V$ so that Q and ε become complex quantities with the real part of ε varying from ε_i to ε_s, and

$$\varepsilon = \varepsilon' - j\varepsilon'' \dotfill 4.4$$

the real and imaginary parts being

$$\varepsilon' = \varepsilon_i + (\varepsilon_s - \varepsilon_i)/(1 + \omega^2\tau^2) \quad . \quad . \quad . \quad . \quad 4.5$$
$$\varepsilon'' = (\varepsilon_s - \varepsilon_i)\omega\tau/(1 + \omega^2\tau^2) \quad . \quad . \quad . \quad . \quad 4.6$$

A circuit analogue to the transformation from eqn. 4.3 to eqn. 4.4 is given in the Appendix. The similarity to eqn. 2.3 results from the parallel assumptions made. Figs. 15a and b give plots of the components ε' and ε'' from eqns. 4.5 and 4.6.

The region where ε' is varying and ε'' is not zero is often called the dispersion region, or the Debye dispersion. Eqns. 4.4–4.6 were derived by Debye (1929), originally from a model representing a

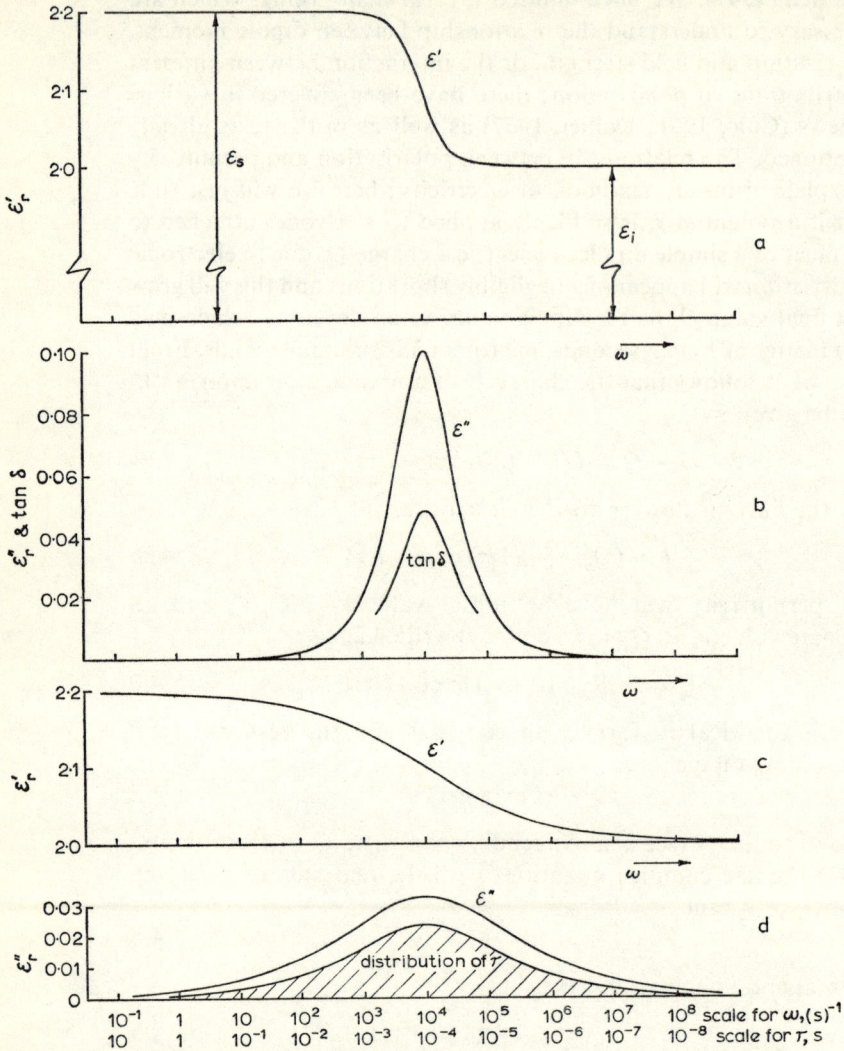

Fig. 15 Graphs of typical quantities in eq. 4.5 and 4.6

    (*a*) Real component ϵ' of permittivity

    (*b*) Imaginary component or loss factor ϵ'' and loss tangent or dissipation factor $\tan \delta = \epsilon''/\epsilon'$
        (δ is the loss angle, $\sin \delta$ the power factor)

    (*c*) and (*d*) ϵ' and ϵ'' for the same total dipole polarisation as in (*a*) and (*b*), but the time
        constants being scattered about the central value with a frequency distribution of shape
        indicated by the shaded area in (*d*)

Numerical values are those of relative permittivity, they are realistic and mutually consistent
but do not refer to any actual dielectric

small dipolar molecule in solution in non-polar solvent; in that model τ was deduced from the notional viscous drag on a sphere the size of the molecule; in some liquids agreement is good. The dispersion region in these cases is usually between 10^7 and 10^{10} Hz.

There is a considerable number of solid systems which match eqns. 4.4–4.6 reasonably well; these are usually not polymeric systems, and the density of dipoles is low enough to avoid electrostatic interaction between them (see, for example, Meakins, 1961). These are rarely materials in general use as insulators; as with

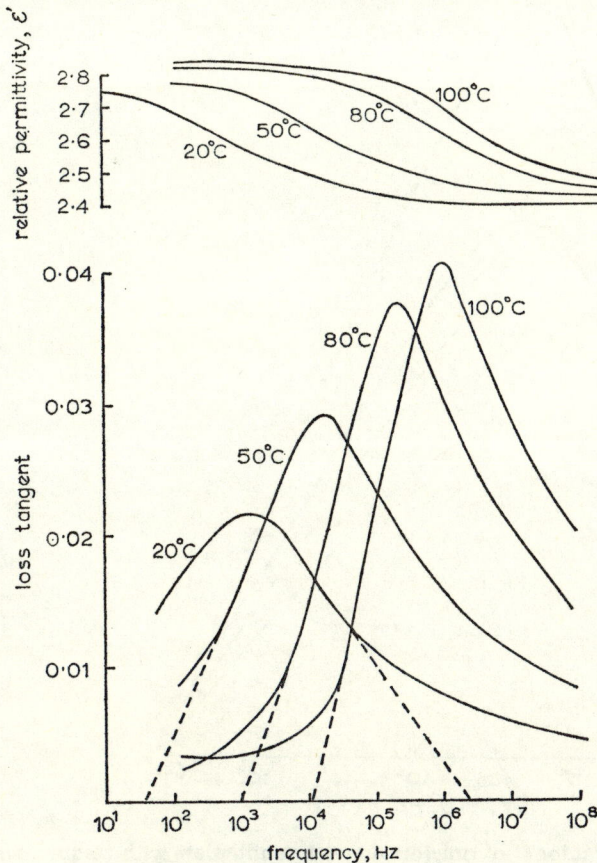

Fig. 16 Variation of relative permittivity ε' and loss tangent $\varepsilon''/\varepsilon'$ with frequency and temperature for polychlortrifluorethylene
(Hartshorn et al., 1953)

mechanical relaxation the behaviour of practical materials usually indicates that the dipoles have a fairly wide distribution of relaxation times, due either to the varying molecular surroundings of the dipoles or to the electrostatic interactions between them, or both. Figs. 15c and d show how the 'ideal' curves of Figs. 15a and b are modified by the spread of relaxation times indicated by the shaded area in Fig. 15d.

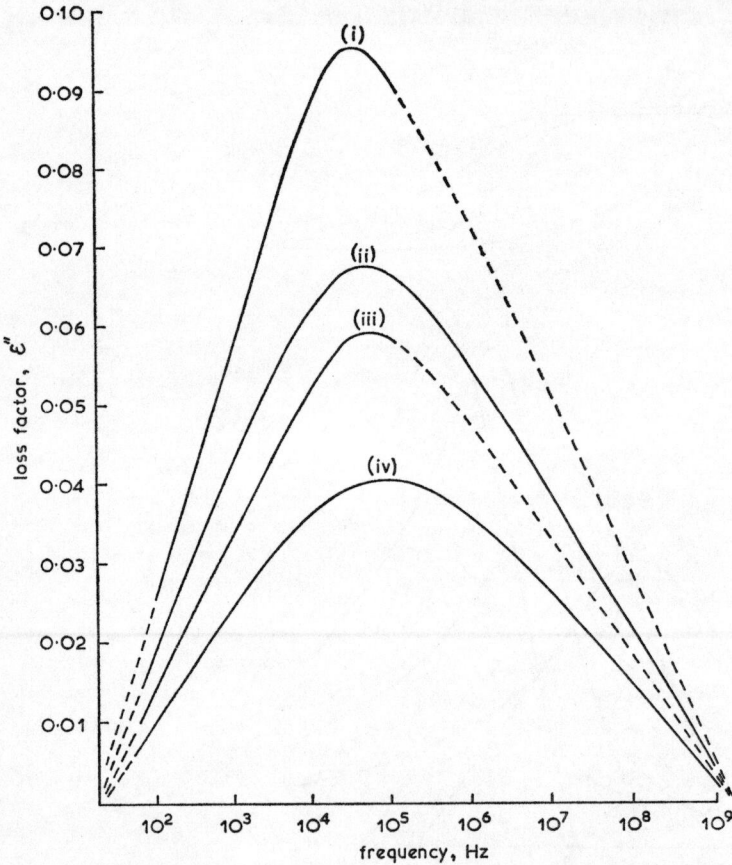

Fig. 17 Variation of loss factor ε'' of poly(ethylene) terephthalate with frequency at 0°C

(Reddish, 1950)

(i) amorphous, wet   (iii) crystalline, wet
(ii) amorphous, dry   (iv) crystalline, dry

Fig. 16 shows results for polychlortrifluorethylene over the same frequency range as Fig. 15, these results are reproduced because of the unusually wide range of measurement. The breadth of the curves indicates a distribution of relaxation times over a range of more than 100:1. Note that as the loss curve becomes narrower at the higher temperatures, (due to convergence of the time constants) it also becomes higher. This is because $\varepsilon_s - \varepsilon_i$ in eqn. 4.4 is dependent on the total dipole moment, and only slightly dependent on temperature; so the total change of permittivity from low to high frequency, likewise the total area under the ε'' vs $\log\omega$ curve remains approximately the same. Fig. 17 shows a plot of ε'' for the material from which Terylene is made, the losses here are attributed to dipole rotation partly of the residual $-OH$ groups at the ends of the terephthalate chain and partly of adsorbed water molecules. Fig. 18 shows the very small loss angles associated with a number of low loss polymers. For reasons similar to those on page 23, $\log\tau$ will be approximately proportional to $1/T$ and roughly, therefore, to temperature itself, so that plots of ε', ε'' and $\tan\delta$ against temperature generally look approximately like reversed plots against log(frequency). Fig. 19 corresponds in this way with Fig. 18. Sets of curves of ε' and $\tan\delta$ against temperature and frequency for a series of dipolar substances of different relaxation times are shown in Figs. 51 and 52 in Chapter 10. These are typical of organic liquids and glasses with a high concentration of dipole molecules. Plots against temperature are much easier to obtain than plots against frequency, and they abound in the literature.

Obviously to get a complete picture we need to have both temperature and frequency variation; this can be displayed on a 'contour map' with temperature along one horizontal axis and frequency along the other, the contour 'height' being permittivity or loss. Relief models made from such maps have been constructed by Reddish (1950) and are shown in Figs. 20 and 21 (Plates). (Fig. 17 is simply a section along the 0°C line of Fig. 20.) Two ridges stretch from a high frequency high temperature corner of Fig. 20, following the same course as the region of maximum slope of Fig. 21, these correspond to two groups of dipoles having different rates of change of τ with temperature.

At one time it seemed difficult to believe that molecular units individually subject to thermal vibration at fequencies of the order of $10^{13}-10^{14}$ Hz could be collectively responsible for losses at 50 Hz, much less for the 'absorption' currents observed when a steady voltage is applied for several hours. It is now clear, with the

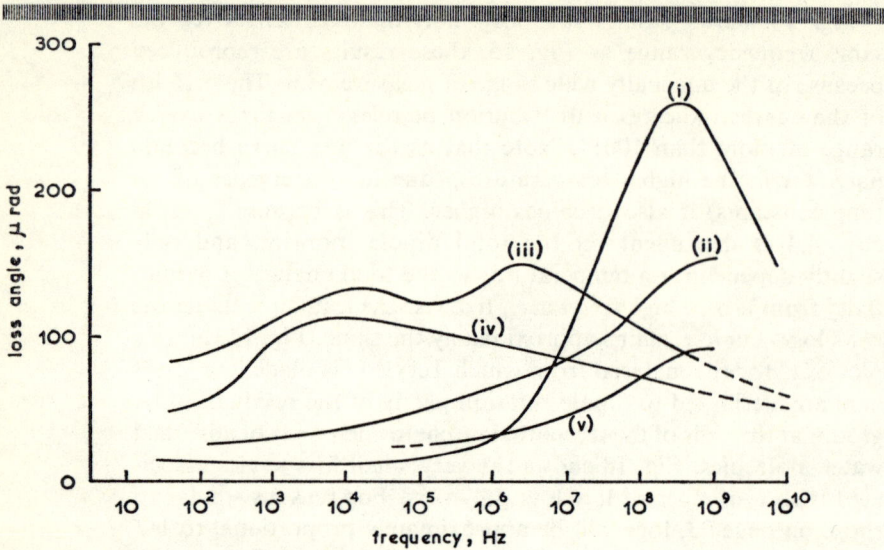

Fig. 18 Variation of loss angle with frequency for some low-loss polymers at room temperature

(Buckingham and Reddish, 1967)

(i) Polytetrafluorethylene
(ii) Polyethylene density 0·923 g/ml
(iii) Propylene-ethylene copolymer
(iv) Polypropylene
(v) Polyethylene density 0·934 g/ml

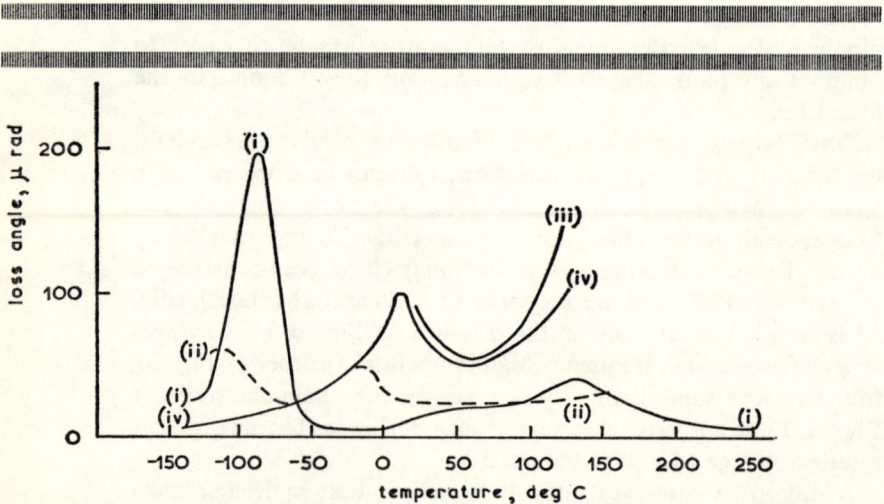

Fig. 19 Variation of loss angle with temperature at 178 Hz for the polymers in Fig. 18

(Buckingham and Reddish, 1967)

Key as for Fig. 18, except higher density polythene omitted

aid of temperature-frequency contour maps, that dispersion regions can be followed from the megacycle range to periods of hours over a temperature drop of the order of 150°C. There seems no reason to exclude the possibility of very slow dipole relaxations at any temperature, at least so long as it is well below the melting temperature of the substance. Many long-time phenomena previously attributed to movements of ions may in fact be due to dipolar polarisation.

Correspondences between electrical and mechanical relaxations times have been sought and found in a number of polymers including Terylene, poly(methyl methacrylate) (Reddish, 1962) and SBR rubbers (Payne, 1959; see also Section 8.14.1), but it does not seem that the one is always accompanied by the other. Either type of relaxation region may be spoken of as a 'transition' implying that it corresponds to a second-order change in structure such as the onset of rotation of a particular type of group. These are often labelled α, β etc., starting with the one at the highest temperature. Remembering the dependence on frequency as well as temperature it will be realised that these transitions are not necessarily sharp.

The effects of plasticisers on dipole rotation are similar to their effects on mechanical properties (Section 2.8). This is illustrated for poly(vinyl chloride) in Section 8.6 and Fig. 43.

4.1.1   Dielectric relaxation in general

Eqns. 4.1–4.6 were derived from considering a particular model of a particular type of polarisation—dipole rotation—but the essential assumptions made were that the ultimate polarisation produced by a constant field is proportional to the field producing it, and that the rate of growth of polarisation is proportional to the difference between the ultimate polarisation and the polarisation already produced, for each individual polarisable element. These assumptions, together with the assumption that the effects of changes of field at various times on one element can be superposed and that the polarisations of many elements of various relaxation times can be summed, follow from the assumption of linearity to response; they were proposed a century ago (Hopkinson, 1876) and seem likely to apply to other mechanisms than dipole rotation, for instance to polarisation by ions translating between neighbouring positions of equilibrium (Section 4.2.3.2). Also it is obvious (see Appendix) that they apply to a resistance–capacitance system and to the assemblage of resistance–capacitance systems which represents a mixture of materials of different conductivities (Maxwell–Wagner dielectric, Section 4.2.3.1).

Consistency with eqns. 4.1–4.6, therefore, implies only that we are dealing with a linear polarisation mechanism; were it not linear the fact should already have been known from dependence on field strength, so that this consistency tells us very little. A study of the effects of temperature and composition on the curves of loss vs frequency and of permittivity vs frequency, however, may give useful clues to the nature of the polarisation through the effects of these on the relaxation spectrum.

Methods of analysing experimental curves to obtain the corresponding distribution of relaxation times have been devised (Smyth, 1955; Cole, 1961; Daniel, 1967) but these distributions do not in themselves usually throw much light on the physical nature of the polarisation. The related methods of transforming from the d.c. to the a.c. relationships, which is more complex than the derivation of eqns. 4.4–4.6 from eqn. 4.3 when there are many relaxation times (Daniel, 1967), provides a means of extending a.c. data to frequencies as low as 10^{-4} Hz without the enormously cumbrous equipment which would be needed to make such measurements directly. Recently it has been found that these methods can supersede the a.c. methods for frequencies up to 1 MHz (Hyde, 1970; see also Section 6.3.4).

It is sometimes useful to note that a distribution, such that the magnitude of polarisation (e.g. number of dipoles) having relaxation times between τ_1 and τ_2 is proportional to $(\log\tau_1 - \log\tau_2)$ whatever the values of τ_1 and τ_2, produces a constant value of ε'' independent of ω and a value of ε' falling linearly with $\log\omega$. The current due to the application of a steady voltage to a dielectric with this distribution is inversely proportional to the time of application. This situation strictly interpreted is of course impossible because it implies an infinite polarisation, but the concept is useful because loss tangents which are nearly constant over a wide frequency range, and hyperbola-like current-time curves are often encountered (especially in glasses, ceramics and materials containing cellulose); in these cases the distribution function plotted against $\log\tau$ is approximately a horizontal line over the corresponding range.

A current/time relationship of the form

$$I \propto t^{-m}$$

m being in the region 0·7–0·9, is often used; this was casually noted by Hopkinson (1897) and has been rediscovered from time to time ever since. The corresponding a.c. loss relationship is

$$\varepsilon'' \propto \omega^{(1-m)}$$

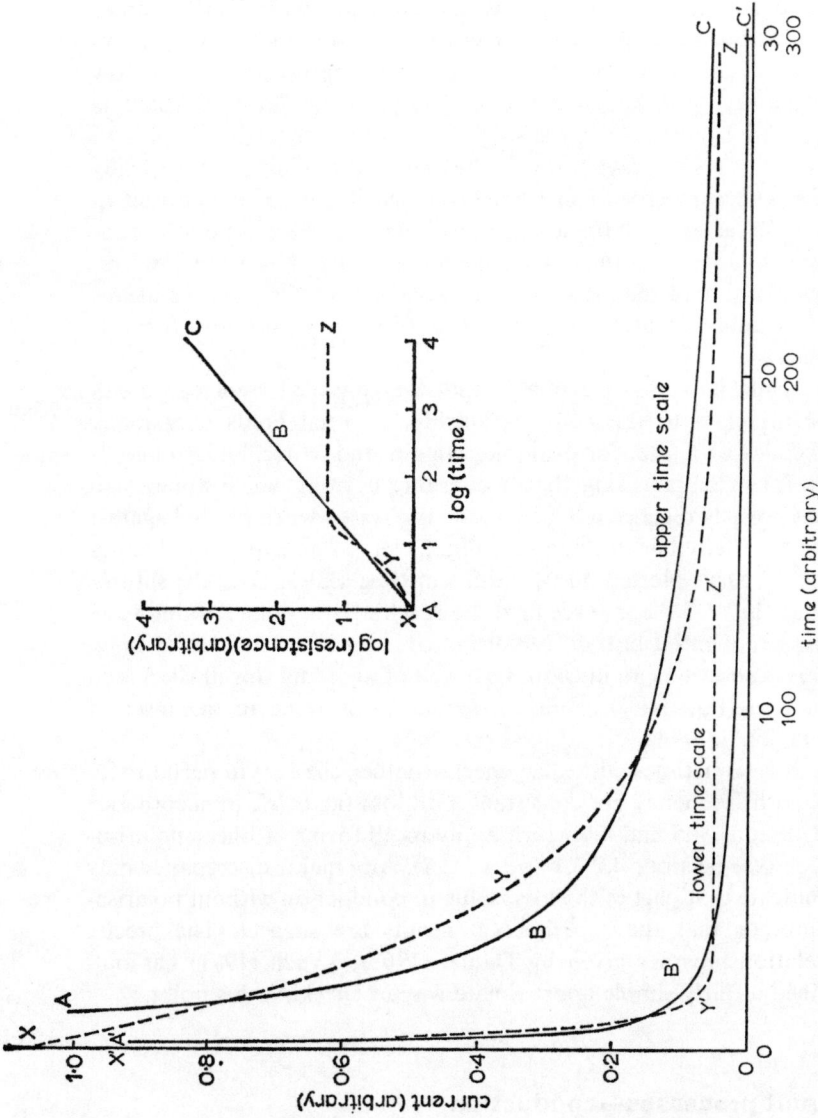

Fig. 22 Full lines A B C and A′ B′ C′: typical curve of charging current against time ($I \propto t^{-0.85}$) for common materials Dotted lines X Y Z and X′ Y′ Z′: curve of exponential time decay to a steady current [$I = \exp(-0.25t) + 0.05$] Each curve is plotted linearly to two different time scales in the ratio of 10:1. Note that on the shorter time scale the power law curve appears to be reaching an asymptote like the exponential, which the longer time scale shows to be illusory Inset: logarithmic plots of inverse of current, or resistance, against log(time) for the same relationships shows very clearly which curve is reaching an asymptote

(see Daniel, 1967) representing a loss factor slowly falling with rising frequency, also common in the materials just mentioned. It follows that when ε'' is rising with rising frequency, i.e. on the low frequency side of a loss peak associated with a wide distribution of relaxation times, the value of m for a log–log plot of current vs time at constant voltage, over periods appropriate to the low frequency, will be greater than unity. A similar conclusion is reached by considering a single relaxation; the log–log plot of $\exp(-t/\tau)$ is concave towards the origin and of slope numerically greater than unity when $t > \tau$. More specifically, if the current at time $2t'$ after applying a constant voltage is less than half that at time t_i, then the angular frequency $\omega = 1/t'$ is on the low frequency side of the peak of ε''. This seems to offer a simple means of roughly locating a peak at a frequency too low for a.c. measurement.

While it seems improbable that any physical significance can be attached to the t^{-m} relationship, it often holds over many decades of t (see, for example, Higgin and Hirsch, 1970) and is useful in emphasising that a charging current which appears to have nearly reached a 'true conduction' value when plotted against time, usually has not done so. Fig. 22 shows an exponential and a $t^{-0.85}$ curve plotted on two different time-scales. Over the shorter time the $t^{-0.85}$ curve seems to be reaching the same asymptote as the exponential but, on replotting over a scale ten times as long, this is shown as an illusion. One way of avoiding this illusion is to plot log(resistivity) against log(time) as is done in the inset of Fig. 22.

It is sometimes wished to check whether changes in permittivity ε' with frequency are consistent with loss factor ε'', in accordance with eqns. 4.5 and 4.6 which apply to all forms of linear polarisation (see Sections 4.2.3.1 and 4.2.3.2). Any major discrepancy may indicate that part of the loss is due to conduction without polarisation, or that the experimental results are suspect. The precise relationships are given by Daniel (1967); Lynch (1971) has outlined a fairly simple approximate way of checking this point.

4.2 Transport processes—conduction

It will be obvious from the previous section that small charging (or discharing) currents due to orientation of slow moving dipoles would be impossible to distinguish from currents due to the movements of ions and electrons which, for reasons to be discussed later, may also decay with time. Nor is there any clear-cut way in which a 'true' charging current can be recognised from an 'anomalous' or 'after effect' current; the current/time curve is continuous from the shortest measurable time after application of a field to the longest.

There is no electrical distinction between currents carried by ions moving bodily through the material and currents due to dipolar polarisation having a longer time-constant than the duration of the measurement. The bodily transport of matter by the former means requires more sensitive methods of detection than have yet been devised. Consequently when we speak of conduction in insulators, which almost always decays with time, though we usually have in mind a transport process which we think is a more likely explanation than a polarisation process, we can rarely test whether we are correct.

4.2.1 Conduction at low stresses

In Fig. 20 the high-temperature low-frequency corner shows a region of rapidly rising tan δ or ε'', which is nearly proportional to the reciprocal of the frequency. Polyvinyl chloride plasticised with dioctyl phthalate shows a similar region of the loss factor (Reddish, 1962) and it is reasonable to conclude that transport conduction is taking place. (The conductivity for dry Terylene (Fig. 20) is of the order of 10^{-9} S/m at 200°C; for the plasticised p.v.c. it is of the order of 10^{-12} S/m at 100°C).

Apart from a few such cases it seems improbable that conduction by ions or electrons accounts for the low voltage properties of dry polymers in so far as they are of practical interest (the effects of moisture are discussed in Chapter 5).

In certain glasses it seems clear that simple electrolytic conduction occurs at room temperature and above (Sutton, 1960) but the glasses which have been most fully investigated are those of high alkali content and relatively high conductivity. In these cases the variation of conductivity with temperature can usually be expressed as

$$\sigma = A \exp(-b/T)$$

A and b being empirical constants. Ceramics develop a significant conductivity at high temperatures, it may be either electronic or ionic depending on the structure of the material; the temperature at which conductivity becomes important may be as high as 1300°C (for alumina).

C

Surface conductivity, despite its considerable practical importance in d.c. and low frequency measurements, is little understood and allows few generalisations. It is greatly increased by increasing the ambient humidity and by the presence of traces of salts, acids etc., in the material or on its surface. It can be reduced by keeping the surface a few degrees above ambient temperature or by treating with a water repellent.

Conduction in liquids at low stresses is hardly better understood. It may be attributed to ions but the ions have not been identified nor have precise properties been associated with them. Mobilities have been measured by various methods (Adamczewski, 1969); looking at the results over a wide range of hydrocarbons mobility is roughly proportional to the inverse of viscosity, varying from about $10^{-9}\,m^2/Vs$ for paraffins similar to transformer oil to $10^{-7}\,m^2/Vs$ for liquids such as n-hexane. This is an overall picture, since mobility is not always inversely proportional to viscosity for one and the same liquid at different temperatures, nor for different members of the same family of hydrocarbons. Agreement between different methods of measurement is not good. Often, but not always, the negative ion is found to be more mobile than the positive and there is evidence that in some liquids there are two or more species of the same sign but different mobility. Thus in hexane or pentane into which electrons have been injected at high voltage the drift mobilities of negative ions at low fields may be of the order of 10^{-3}–$10^{-5}\,m^2/Vs$ (Dublin, 1972). Either, or both, of these may represent electrons which 'hop' from molecule to molecule with pauses between the hops rather than electrons permanently attached to molecules or groups of molecules. It may be that these high mobilities are the result of measurements taken without sufficient regard to the bodily movements of the liquid resulting from the high voltage injection (Taylor and House, 1972; Hewish and Brignell, 1972).

Conduction in transformer oil is increased by moisture but there is some evidence that very pure hydrocarbon oil exposed to humid air does not show any conduction. Experiments with purified liquids, including argon and other liquefied gases, have produced many observations about the effects of electrode preparation, dissolved impurities, and other circumstances, together with estimates of mobility varying from 10^{-7}–$10^{-11}\,m^2/Vs$ (Lewis, 1959).

Another interesting aspect of conduction in liquids has appeared from efforts to development high voltage electrostatic generating machines using liquids of high permittivity such as nitrobenzene.

Normal purification techniques do not produce a higher resistivity than about 10^6–$10^7\,\Omega$m in nitrobenzene (compared with up to $10^{17}\,\Omega$m for a very pure hydrocarbon). It is not clear whether this is wholly attributable to the fact that ionisable impurities are very much more highly dissociated in nitrobenzene (which has a permittivity half that of water) or whether the nitrobenzene itself provides the ions. It is proved, however, that electrodialysis with ion-exchange resins increases the resistivity to about $10^9\,\Omega$m where it remains for a substantial time. A further increase to 10^{10} or $10^{11}\,\Omega$m can be obtained by using electrodialytic membranes themselves as electrodes (Sauviat and Tobazeon, 1970). The further details of this process are not well enough understood to warrant discussion here. For practical implications see Section 10.7.

4.2.2 Conduction at high stresses

We have already noted that conduction, in the sense of more or less steady currents carried by ions or electrons, is exceptional in solid insulation at low stresses and normal temperatures; but as the stress approaches that of breakdown the current increases exponentially and does not decay nearly so markedly with time, for a steady voltage. Typical cases are described by Taylor and Lewis (1971) and by Bradwell, Cooper and Varlow (1971). This current is generally accepted as being electronic since it must result from injection of carriers from an electrode or multiplication in the material, or both, and electrons are the only carriers expected to behave in this way. The highest stresses at which measurements were made by the authors just mentioned were a factor of three or four below the (discharge-free) breakdown strength, the resulting current being of the order of $10^{-10}\,\mathrm{A/cm^2}$ (polyethylene terephthalate) to $10^{-7}\,\mathrm{A/cm^2}$ (polyethylene).

The exponential current/voltage relationship is generally attributed to the mechanism of emission of electrons from the cathode into the insulation, with two main alternative explanations. One is that the electrons pass over a barrier whose height is lowered by the applied field, the number of electrons having sufficient energy increases very rapidly as the required energy decreases (Schottky emission). The other is that field emission (tunnelling) takes place through a thin barrier whose transparency is influenced by the field. Electron and ion multiplication analogous to that in a gas discharge may also occur at the higher stresses. The space-charge of the injected electrons (whether moving or trapped) within the insulator will have the effect of decreasing the field at the cathode and modifying the current/voltage relation.

Unfortunately the available knowledge of dielectrics only allows us to form a solid-state model which is as crude as the corresponding models of semiconductors were 35 years ago, so that discussion of their electrical behaviour is as inconclusive as it was for semi-conductors at that time.

Where liquids are concerned similar phenomena appear and have been given similar explanations, but there are at least two other possible means of charge transfer to be considered. First the well known tendency of oil to display movement when under stress suggests that charges injected into the liquid from an electrode attaching themselves to oil molecules in the neighbourhood cause this oil to be attracted to the opposite electrode where some of its charge may be given up, thus producing a convection of charge independent of conductivity. These movements have been observed by many authors (e.g. Hewish and Brignell, 1972). Gray and Lewis (1968) using a streak of dye found liquid velocities comparable with the expected ion velocities at stresses of the order of 100V/mm. Charge convection in nitrobenzene has been intensively studied by a group at Centre Nationale de la Recherche Scientifique at Grenoble (for example several authors at Dublin, 1972) in connection with electrostatic generation methods.

The second possibility is that particles, which are often observed moving to and fro between the electrodes if a highly stressed region is examined with a microscope, will ferry charge from one electrode to the other. If these particles are assumed to be conducting both the charge they acquire when they touch an electrode, and the velocity with which they travel under the influence of the field can be calculated (Krasucki, 1967a). This could provide an appreciable current at high stress. In most modern experiments on liquids visible particles have been eliminated, and while sub-microscopic particles could account for some observed currents a quantitative check with the theory is impossible at present. It can be shown that many phenomena of conduction in apparently pure liquids can be plausibly attributed to particle movements without any unreasonable assumptions (Birlasekaran and Darveniza, 1972; Krasucki, 1972), but there is not yet clear evidence that this mechanism plays more than a minor part in normal circumstances (Rhodes and Brignell, 1972). It has been observed that centrifuging of purified cyclohexane reduces its conduction to a low value which is nearly time-independent (Matsuzewski *et al.*, 1972).

A normal feature of conduction at high stresses is the occurrence of irregular pulses of duration of the order of microseconds. In solids these are usually attributed to discharges at electrode edges, gaps between electrode and dielectric, or faults in the dielectric. In the cases of either liquids or solids near breakdown stress they are likely to be symptoms of abortive breakdown processes and attributable to the same mechanism as breakdown itself. Liquids often display current pulses which cannot plausibly be attributed to either of these causes and may represent the near approach or arrival of particles at one or other electrode.

4.2.3   Conduction decaying with time

In Sections 4.2.1 and 4.2.2 (apart from the last paragraph) we have considered phenomena which display a substantially steady current and are similar to conduction in electrolytes or semiconductors.

We have already emphasised that such phenomena are the exception rather than the rule, a fact which has often been obscured by experimenters who adopt some arbitrary definition of conduction (such as the value a given length of time after application of voltage) and then discuss the result as though it were indeed simple conduction.

We have referred to the empirical t^{-m} law in Section 4.1.1; there is no particular reason why it should not result from ionic or even electronic conduction; its chief significance is probably that any complex combination of relationships between two variables, without discontinuities or sharp curvatures, is likely to look like a roughly straight line over a limited range when plotted on log–log graph paper.

There are a number of known and conjectured reasons why ion and electron transport should decay from the moment of applying a direct voltage, or display an a.c. power loss which does not vary inversely with frequency. These are considered in the next three sections. (Note that the term ion is used in the same sense as an ion in an electrolyte, it does not imply a gaseous discharge unless the context so indicates.)

4.2.3.1   Heterogeneity

A solid whose conductivity is not uniform will in general show a current decaying with time. More simply, a solid composed of two materials, one of which conducts appreciably while the other does not, the conducting phase being in discrete, or largely discrete, particles embedded in non-conducting phase, or the two forming a sandwich, will display a permittivity and loss angle obeying equations which can be written in a form identical with eqns. 4.2–4.6.

This is known as a Maxwell–Wagner dielectric, and the quantitative relations are given by Daniel (1967, Chapter 14); more general theories of two-phase heterogeneous dielectric are summarised by van Beek (1967) from the papers which he cites. Briefly the dispersion and loss curves, similar to those of Fig. 15, appear at a

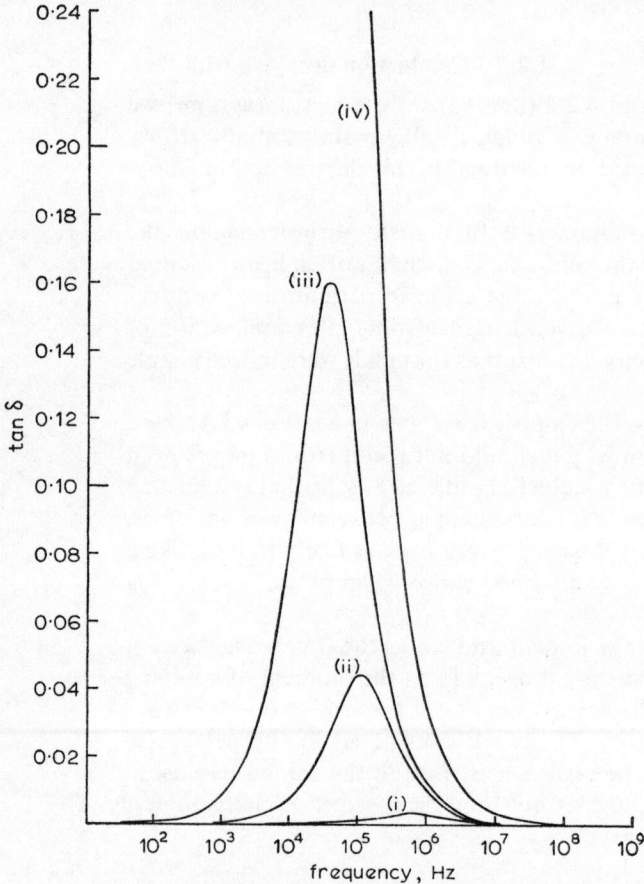

Fig. 23 Calculated loss tangent of composite dielectric composed of a good insulator of relative permittivity 5 containing 2 volumes per 1000 of water of conductivity 5×10^{-3} S/m disposed as

(i) needles of length to diameter ratio 5:1
(ii) needles of length to diameter ratio 20:1
(iii) needles of length to diameter ratio 40:1
(iv) tubes parallel to the field extending through the dielectric

In cases (i), (ii) and (iii), approximate allowance has been made for random orientation of the needles. The same quantity of water disposed as spheres (Wagner's model) would produce a loss tangent maximum of 0·0005 at 1 MHz

point along the log(frequency) axis determined by the conductivity of the conducting phase, the permittivities of both phases, and the shape of the conducting particles. If there is more than one conductivity or shape there will be a distribution of relaxation times. It is important to know that the magnitude of the loss peak and of the permittivity 'step', though independent of the conductivity of the conducting phase, is not only proportional to the quantity of conducting phase but increases as the particles become elongated along the direction of the field, the relaxation time being increased also. To illustrate this Fig. 23, curves (i), (ii) and (iii), show the calculated loss resulting from a small proportion of water in long channels in a good insulator. The longer the channels in relation to their diameter, the greater the loss produced by a given quantity of water, and the lower the frequency of the loss peak for a given conductivity of water. It is thus possible to have a high loss with as small a quantity of water as one chooses at any frequency provided it can be in the form of sufficiently elongate needles with sufficient conductivity. The only limitation is the one illustrated by Fig. 23, the loss at any frequency cannot be greater than would be produced by the same quantity of conducting material of the same conductivity in channels right through the dielectric in parallel with its capacitance.

The application of this to damp fibrous materials or to solids containing cracks and fissures is obvious. It would be quite possible to construct a dielectric containing less than 1 % of water having a tan δ of say 0·05 at all frequencies from a few megahertz to zero (Sillars, 1937). Equally a solid containing 5% of water all in the form of spherical droplets need not have a measurable loss up to at least the megahertz range.

4.2.3.2 Ions with limited travel

We have already considered (Sections 2.3 and 4.1) the way in which organic groups have a probability of moving or rotating past their neighbours depending on the amount of energy that is required to enable them to overcome the 'potential barrier' which is the formal representation of the forces obstructing these movements. Similarly an ion in an inorganic crystal rests in a potential 'well' formed by the attractions and repulsions of its neighbours. An ion in a normal position in a regular crystal of one of the types used for ceramics, has such a minute chance of moving at ordinary temperatures that it need not concern us, but an ion in an abnormal position in a regular crystal, or an ion in a glass (which has not got a regular structure) may be in a much shallower potential well, of depth such that it will jump to another similar well with some frequency in the range of electrical interest, that is to say its relaxation time at normal temperature is something between about 10^{-11}s and, say, 10^3s. In Fig. 24 the full line represents a double potential well, a positive ion lying in one half will oscillate about the minimum, and whenever it happens to acquire enough thermal energy it may jump to the other half. If an electric field is applied, the potential energy diagram will be altered in the way indicated

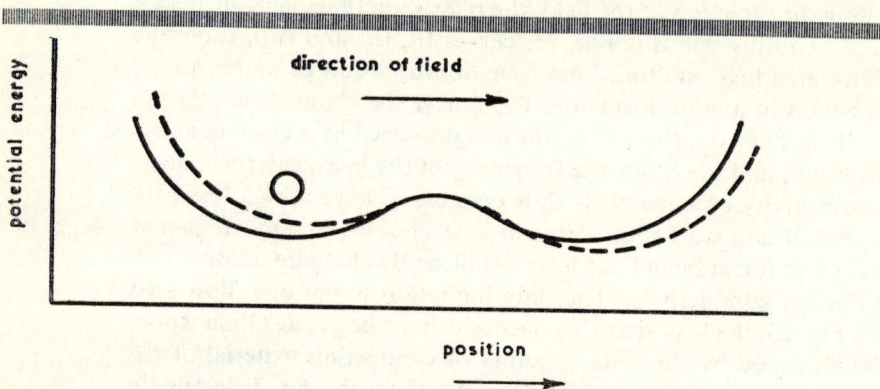

Fig. 24   Potential well model of an ion with two stable positions
Full line without applied field, dotted line with applied field

by the dotted line, with two results. First, the minima in both halves of the well are displaced so that the centre of oscillation moves in the direction of the field. Thus all such ions on the average move with respect to the rest of the structure and a polarisation occurs within a few oscillations, a time short compared with any we are interested in. (This applies to all ions whether in a double well or a single one.) Second, the minimum in the 'downfield' side will be a little lower than that in the 'upfield' side, and so this ion will spend more time in the forme (Daniel, 1967). Looking at the dielectric as a whole there will be more positive ions in the 'downfield' wells, this represents a polarisation which develops exponentially with a time constant determined by the probability of jumping between the two wells. This situation is analogous to that of the dipole moving between two stable positions of orientation and leads to eqns. 4.2–4.6 in the same way (provided all the wells are identical); the relaxation times will decrease roughly exponentially with temperature for reasons similar to those applying to dipoles. Obviously elaborations of this situation with different depths of wells giving a spread of relaxation times can be plausibly supposed. A string of shallow wells along which the ions could move might be supposed to exist, and the resulting losses would be spread over the frequencies appropriate to the most probable times of dwelling in these wells, and combinations of these.

Losses of this kind have been much discussed in connection with glasses (Stevels, 1957; Sutton, 1960, Isard, 1962), under various names such as 'dipole', 'deformation' and 'migration' losses, the lines of demarcation between these being rather indefinite. (In this context 'dipole' does not mean a rotating molecular dipole, but the type of polarisation we have just described.) It is probable that some of the losses in ceramics are of a similar nature.

A related idea, having something of both dipole and ionic concepts, put forward by Meakins and Sack, envisages in certain materials a chain of hydroxyl groups linked together by hydrogen bonds; rotation of all the groups in the chain effectively means the transfer of a hydrogen ion from one end to the other. This is summarised by Daniel 1967 (Chapter 15) who also discusses its possible relevance to the behaviour of damp paper.

Limited travel of ions on a macroscopic scale, as between layers of solid film in impregnated capacitors, or between electrodes at which they are not discharged ('Garton effect') is considered in Section 4.2.3.3.

4.2.3.3 Electrode and barrier effects with ionic conductivity

Ions in insulating solids and liquids are in very different circumstances from those in electrolyte solutions, though their situation is broadly similar. They are very sparse, they have lower mobilities, and the rate of dissociation and recombination appears to be very slow, if some are removed the equilibrium is not restored quickly. Consequently, if, say, positive ions leave the area of the anode they may not be replaced by dissociation and they cannot be supplied by an anode which does not contain them. The current will decrease with time and generally a space charge of negative ions near the anode will develop. Such situations probably arise in oils, they are known to occur in measurements on soda glasses unless electrodes of sodium amalgam or sodium slats are used. Moreover, a solid dielectric, unless the electrodes are coated on to it, will not be in sufficiently intimate contact to allow ions to discharge except at a few points, so that ions will accumulate without being discharged. Space charges in dipolar liquids have been observed directly by optical means (Kerr effect) (e.g. Durand and Fournié, 1970, Pollard and House, 1972, Cherney and Cross, 1973), and by means of probes (Forster, 1967). These effects will appear in d.c. and low

C*

frequency a.c. conditions, at higher frequencies the full ionic conductivity will take effect.

There is evidence, surprisingly at first sight, that even in a liquid dielectric the ions, which are presumably charged organic molecules, or aggregate of molecules, do not necessarily discharge on reaching the electrode to which they are travelling. Once they have reached this electrode they cannot contribute further to conduction unless the field is reversed.

Consequently if an alternating field sufficiently strong to carry one species of ion from one electrode to another in one half-cycle is applied, and if dissociation and recombination in one half-cycle are negligible, a situation is reached in which that species of ion contributes to conduction for a smaller proportion of each half-cycle the higher the amplitude of the field, and the loss will diminish with *increasing* stress. The same result occurs when the motion of the ions is restricted by any other barrier, as when ions move in the narrow space between two layers of paper or plastic in an impregnated capacitor. This, the 'Garton effect', has been observed in thin layers of oil-impregnated paper (Garton, 1941; Daniel, 1967) in good accord with theoretical predictions.

While most of the effects mentioned (and others not mentioned) in this section are in principle calculable, we do not (apart from occasional exceptions like the Garton effect) know enough about the numerical values involved to decide which are likely to be important and which negligible in given circumstances; while any attempt to do a generalised calculation would lead to an unmanageable result.

It should be added that the effects discussed in this section, involving depletion of ions and space charges, will give non-linear as well as time-dependent effects, so that successive increments of voltage will not produce identical increments of time-decaying current, and the effects cannot simply be added together as was assumed in Section 4.1 for instance.

4.3 Electric breakdown of solids

The ability to withstand a given voltage continuously, perhaps with a superposed impulsive voltage occasionally, is the one basic requirement of most insulation, but in surprisingly few circumstances is its thickness determined by testing a sample of the material for breakdown strength. The thickness of insulation on domestic wiring cable is 10 or 20 times that which a simple test would suggest to be necessary; the spacing between terminals on a board is far larger than necessary to prevent simple flashover at any voltage expected to occur between them, and so forth. Even in high-voltage equipment the thickness of insulation must be much greater than would seem necessary from the 'electric strength' as normally measured and quoted.

The reasons for adopting particular types and thickness of insulation are often determined by failure processes which are not related to that which occurs when a high stress is applied; even when high-voltage failure is the critical factor a test lasting a short time does not represent a condition which can be maintained for years. Breakdown strength is not a design quantity therefore, it is a guide for comparing materials and it is extremely useful as a quality control test. Proof tests, though usefully applied to most apparatus and components, are not by themselves any assurance that insulation will give satisfactory service, they merely verify that particular types of gross manufacturing error, or of degradation in service, have not occurred.

The measured breakdown voltages of samples of the same material of different thicknesses do not usually increase in proportion to thickness. Breakdown voltage is influenced by practically every physical factor that can be imagined to be relevant, including temperature, humidity, duration of test, whether d.c. or a.c., frequency if a.c., area of test electrodes, discharges in the ambient medium (usually air or oil) or in cavities within solid insulation, polarity if there is asymmetry of the electrodes, pressure on the electrodes, any recent applications of stress, and many other factors.

The fundamental breakdown processes are not understood; not for lack of experimental observations but because our background knowledge of molecular solids such as polymers is too crude to form more than a very rough picture of the behaviour of energetic electrons within them and of their effects on the structure. Future progress is unlikely to come from still more observations of voltage, thicknesses, temperatures, impulse shapes, probabilities and so forth; it requires new efforts, theoretical and experimental, to distinguish the individual microscopic processes forming the succession of events which we call a breakdown; to develop techniques for investigating each of them, as has been, and is still being done, for the less complex gaseous state.

Three main types of breakdown are generally recognised. The first, a purely laboratory phenomenon, involves the concept of the highest electrical stress that a particular material will stand in the absence of any recognisable secondary effects such as heating by dielectric loss or bombardment by charged particles from extraneous discharge. The second type is the kind of breakdown which is now thought to be the normal one in testing and in service (usually occurring at stresses one or two orders of magnitude lower than the first type) and appears to be initiated and propagated by discharges in air or oil impinging on the solid. The third type is important mainly because its cause can be recognised and usually avoided: it results from self-heating by dielectric loss (or conduction current) to the point at which the solid melts or decomposes into liquids or gases, resulting in breakdown due to the partial disintegration of the original solid. These types are the subject of the sections which follow. They are readily distinguishable on a macroscopic scale, it may be that on a microscopic scale two, even all three, become indistinguishable.

4.3.1 Discharge-free ('intrinsic') breakdown

Just as a particular length of gap between large spheres in air has a breakdown strength which is known and reproducible it is reasonable to expect that a given thickness of solid insulation has definite breakdown characteristics. It is found that the highest breakdown fields are obtained when the sample is so designed that there is a high stress in the centre of the solid under test but too low a stress at the edges to cause discharge in the surrounding medium, as for example the arrangement in Fig. 25. Discharge-free measurements cannot be made on thick specimens, and obviously the specimens must be free of internal cavities.

For some years it was accepted that results thus obtained represented an intrinsic property of the material tested, and that this electric strength was substantially independent of thickness (see Mason, 1959) but it has become increasingly doubtful whether a reproducible breakdown resulting directly from the electric field has been identified (Riddlestone, 1953; McKeown, 1965) or whether breakdown can ever be characterised by a particular field strength. Injection of electrons into a dielectric, if as seems probable it occurs, will produce a negative space charge decreasing the

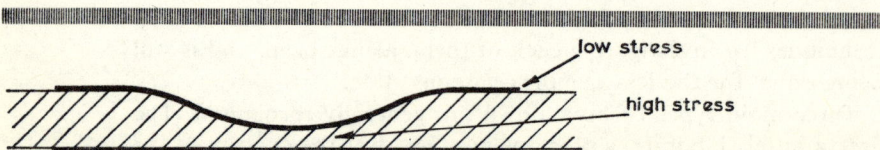

Fig. 25   One type of electrode used for measuring discharge-free breakdown

field near the cathode and increasing it near the anode. Bradwell, Cooper and Varlow (1971) have given experimental evidence of this in polymers, particularly in the fact that impulse strength is greatly decreased by prestressing of the opposite polarity to the impulse.

The breakdown stresses found in discharge-free tests on hydrocarbon polymers lie in the range 0·5–1 MV/mm at room temperature, falling by a factor of order two as the temperature increases to 100°C and rising at lower temperatures to a higher plateau. These tests are on small areas and usually last milliseconds or seconds only. Artbauer and Griac (1970) have investigated the effect of stressed area and duration of test; extrapolating very greatly they conclude that these and the margin of reliability needed in service may account for most of the factor of 100 or so between the strengths mentioned above and the stresses which are practicable under the best conditions of service. There is evidence that discharge-free breakdown is initiated by very small embedded foreign particles.

Several theories of ultimate breakdown strength have been proposed, based on various models of the solid state, neither the experimental data nor the theories themselves can yet be given the precision necessary to favour one theory unequivocally. A review of these is given by Cooper (1966).

The mechanical pressures associated with the high electric stress have been found to influence breakdown strength (Cooper, 1963). Indeed, it seems probable that in soft rubbers breakdown may be the result of electromechanical instability; compression decreases the thickness which increases the stress until at some point, if the elastic modulus is low enough, the process becomes catastrophic (Stark and Garton, 1955; Blok and Le Grand, 1969). This requires either that the electrodes distort with the material, or that sufficient charge can be conveyed across a gap between electrode and dielectric to maintain the voltage.

4.3.2 Progressive breakdown in solids

Except in the rather special circumstances described in the previous section, breakdown in polymers is probably a progressive cumulative process, involving the initiation and growth of one or more tiny tubular fissures (typically 1–10 μm diameter) which may form a tree-like structure. Such trees have been observed in polyethylene cable insulation after service (e.g. Kitchin and Pratt, 1958; Vahlstrom, 1971), their growth has been studied by various workers using needles or sharp-pointed holes to initiate the tree

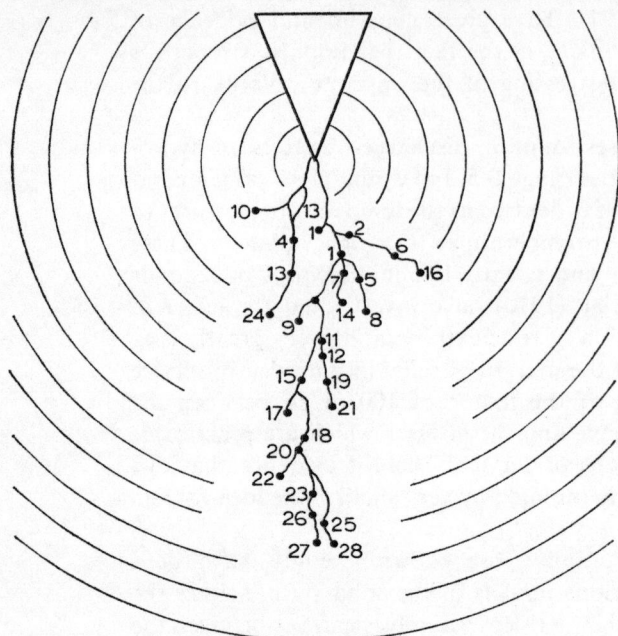

Fig. 26   Typical progress of treeing in poly(methyl methacrylate) from the point electrode at the centre

Points numbered 1, 2, 3 etc. indicate the extent of growth as it appeared after the 1st, 2nd, 3rd etc. discharge

(Bolton, Cooper and Gupta, 1965)

(Olyphant, 1963; McMahon and Perkins, 1964). Bolton, Cooper and Gupta (1965) made a statistical study in Perspex using a needle-shaped hole having an end radius of about $25\,\mu m$, coated with a conducting film, and with its end situated 5 mm from a plane face similarly coated. They applied impulses of some 150 kV amplitude and $30\,\mu s$ duration to this system; each time a visible discharge was seen they noted where the tree system had been extended. Fig. 26 shows a typical result, from such results they concluded that growth was largely random, i.e. the nth extension rarely occurs from the $(n-1)$th and each extension originates from the 'stem' rather than from the tip of a previous extension on almost half the occasions. The tubules were about $1\,\mu m$ in diameter, those near the electrode becoming wider with use. The visible discharges lasted $2-30\,\mu s$, the average extension was about $400\,\mu m$ per discharge but varied by a factor of about seven up and down. When the tip of the tree came near to the plane electrode these habits were modified in favour of growth towards it. At the end of the treeing process the resistivity of the region of the tree was very high. Fig. 27 (Plate) is a photograph of a similar type of tree from a needle embedded in Perspex (Forrest, 1963). Similar conclusions have been reached by other experimenters.

The growth of these breakdown channels evidently depends on a gaseous discharge from the originating electrode to the growing point along a tubule only a few mean free paths in diameter; the nature of the process at the growing point, and the reasons for confinement to such a narrow tunnel but not wholly the pre-existing tunnel are still matters for speculation. General explanations attributing it to stress concentration at the tip of the tubules (which appear to be finely tapered), to volatilisation by bombardment by ions or electrons in the discharge, to decomposition by chemical process excited by the discharge or to rupture by electromechanical forces are plausible but do not account for the pattern of growth. It has been reported that various substances have an inhibiting effect on channel propagation; Billing and Mason (1970) refer to these findings but themselves find that only certain halogenated additives have a marked effect. Treeing in polymer films seems to be discouraged by immersion in chlorinated diphenyl.

4.3.3 Breakdown initiated by extraneous discharges (partial discharge or 'corona' breakdown)

The breakdown which develops when a plane polymer surface is subjected to discharge through the surrounding air usually begins with a general erosion of the surface developing later into a pit of indefinite shape from which, sooner or later, one or more tubules develop. These appear to be similar to those which develop from 'artificial' needle points as already described. It is now clear why, in a normal breakdown test, the breakdown rarely occurs under the electrodes (McMahon and Perkins, 1963).

As would be expected, breakdown develops more rapidly the higher the overvoltage, and different materials display very different degrees of resistance to prolonged discharge. Generally discharges under oil are more damaging than in air; this is attributable to the higher energy associated with each discharge. Cavities within insulation give rise to similar erosion and breakdown if the field strength is sufficient to cause breakdown of the air or gas within them, this is considered in more detail in Section 4.3.5.

At a given alternating voltage, the number and intensity of discharges in a half-cycle is roughly independent of the duration of the half-cycle, i.e. of the frequency. Consequently, if the progress towards breakdown is, on the average, the same for each discharge, doubling the frequency will halve the time to breakdown. This method of accelerating life tests under discharge is limited by the appearance of thermal breakdown at the higher frequencies if the material has appreciable dielectric loss.

Various arrangements of artificially contrived holes in insulation and curved or pointed electrodes on or near insulation surfaces have been used to compare the endurance of different materials to discharges impinging on them (Mason, 1959, and references given there; Mason, 1960 and 1965; Hogg and Walley, 1970). It is usual to plot log(life in cycles) against $\log(V/V_i)$, V being the applied voltage and V_i the lowest voltage at which discharge can be detected in the particular arrangement being used. It is generally found that no deterioration takes place below V_i, above it the time to breakdown is inversely proportional to some fairly high power, such as six or eight, of the voltage. Most of these tests have been carried out at frequencies from 500–2000 Hz and referred to 50 Hz by simple proportion. The validity of this method of accelerating the test (subject to avoiding thermal breakdown) has been checked by direct measurement at 50 Hz (Howard, 1951; Mason, 1960; Hogg and Walley, 1970) but it is still advisable to check this point for material not previously tested.

Reproducible results, comparable at different frequencies, are only obtained if pure dry air is circulated around the discharging electrodes, to remove moisture and acids. Non-removal of these may shorten the life of the specimen. On the other hand, where tests are done on enclosed cavities the discharge may cease altogether for periods due to the water forming a conducting film on the surface of the cavity and shielding it from the field, thus lenthening the life.

There are no generalisations to be made about comparative lives of different organic materials, Fig. 28 gives an idea of relative lives of a few common ones. Other comparisons are given in the articles by Mason already mentioned and by Smyser (1969).

It will be obvious that proof testing at voltages well above the operating voltage of a piece of equipment, if it causes discharge, is likely to produce some progress towards breakdown; though reassuring that the insulation was not at the point of failure such tests may bring failure appreciably nearer. Tests by discharge measurement provide an alternative, in principle, which will be referred to in Chapter 6.

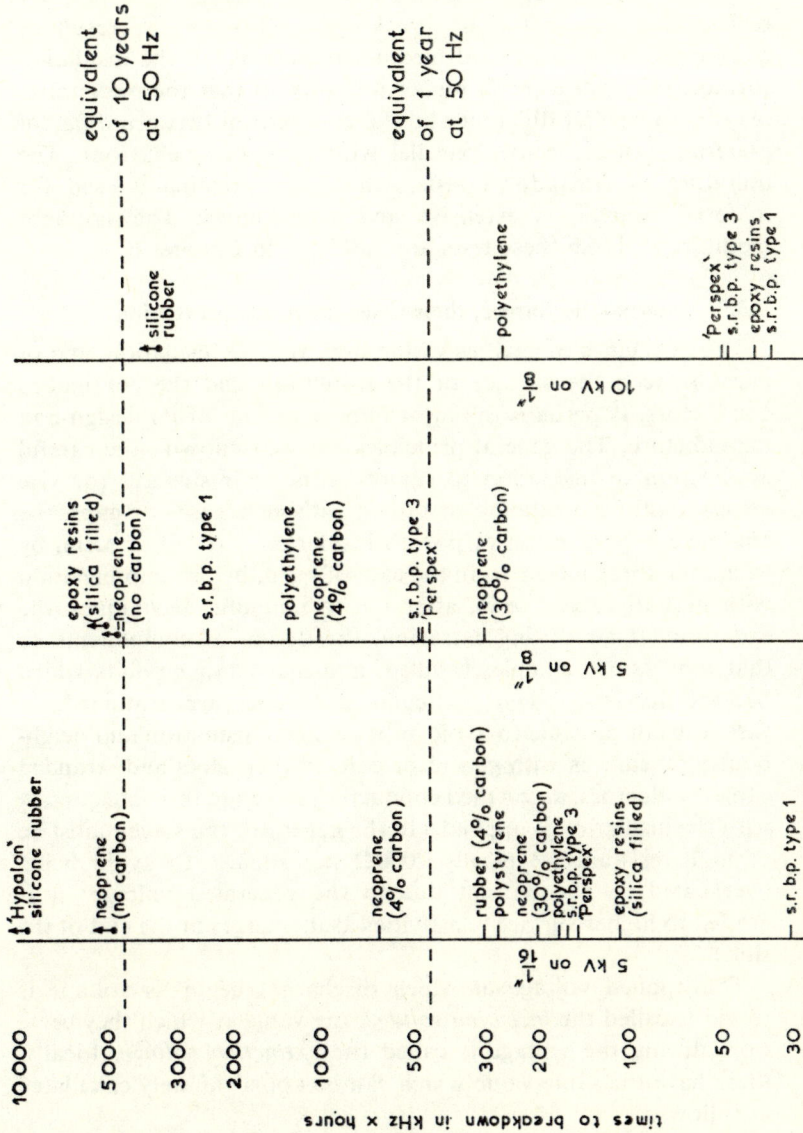

Fig. 28 Diagrammatic representation of relative lives of various materials under surface discharge (Rollinson, 1957) Inorganic materials and resin-bonded mica (but not resin-bonded glass fabric) have much longer lives than the above

4.3.4   Breakdown in electric strength tests

Laboratory tests are usually conducted so as to produce break-down in one minute rather than several years. There is every reason to think that these breakdowns also are the result of discharges in the air or oil medium impinging on the insulation surface, but with a much higher intensity so that the mechanism may be somewhat different and the comparison between different materials not altogether parallel with long-time behaviour. The literature of breakdown tests, which are principally used for material control, is extensive and well known. The standard methods of doing these tests are indicated in Chapter 6.

4.3.5   Internal discharges, their assessment and prevention

The avoidance of cavities within high voltage insulation, and of gaps between the surface of the insulation and the contiguous conductors, is perhaps the most important aim of its design and manufacture. The general principles are well known—the careful application of insulation to smooth conductor surfaces (or vice versa), uniform wrapping or taping without creases or gaps, the complete impregnation of porous insulation with oil or resin, by vacuum impregnation or, in the case of resin, by pre-impregnation with partially cured resin, attention to moulding techniques, the exclusion of air during extrusion, the design of components so that moulds are completely filled, and many other points which become apparent when particular techniques are examined. In case it is not possible to avoid gaps between insulation and neigh-bouring metal, as with generator coils in their slots and stranded cable conductors, an earthed conducting screen in intimate contact with the insulation is applied. (In the generator the screen must be of high resistance, typically 2000Ω per square, to avoid being overheated by the current due to the generated voltage; it is graded to higher surface resistivities as it emerges at the end of the slot.)

The applied voltage at which discharges begin as voltage is raised is called the *inception voltage*, the value at which they cease on reducing the voltage is called the *extinction voltage*. Ideally, these have the same value which can be approximately calculated as follows.

A small cavity in a dielectric subject to an alternating field will have within it a field of magnitude lying between E_D and $\varepsilon_D E_D$ where ε_D is the relative permittivity of the dielectric and E_D is the (assumed uniform) field in the dielectric at a point several diameters of the cavity distant from it. If the cavity is a tunnel parallel to the field, the field within it will be E_D, if a disc perpendicular to the field, $\varepsilon_D E_D$; if the cavity is spherical the field within it will be

$$\frac{3\varepsilon_D}{2\varepsilon_D + 1} E_D$$

If this field reaches the breakdown strength of the air or gas within the cavity discharges will take place. Fig. 29 shows the breakdown strength of air at atmospheric pressure as a function of gap between test electrodes. If it is assumed that discharge between layers of charge on insulating surfaces obeys the same relationship, it is clear that the smaller the cavity the greater the field required to cause discharge. For example, at $20\,kV/mm$ a cavity about $10\,\mu m$ across should not discharge. Measured a.c. inception voltages with cavities of known dimensions usually accord with this kind of analysis, though discharges on impulse voltages seem to be affected by other factors also (Densley and Salvage, 1971). Parkman (1959) has pointed out that the calculated a.c. discharge inception voltage of a dielectric containing a cavity, plotted against the size of the cavity, falls to a minimum which is shallow for cavities in materials of relative permittivity 2 but deep for materials of permittivity 6 or 8; it rises from the minimum as the product (cavity depth)×(relative permittivity) becomes comparable with the thickness of the insulation. Thus, not only is the inception

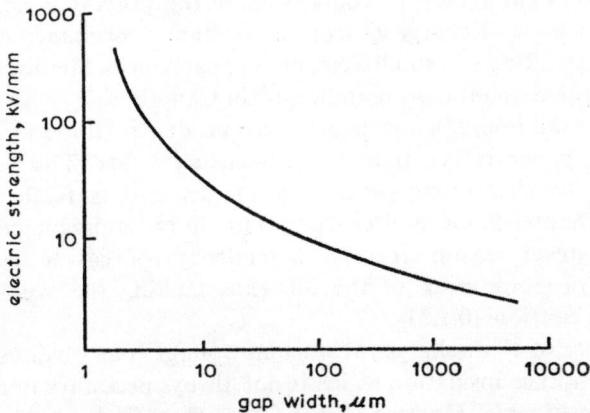

Fig. 29 Electric strength of air at atmospheric pressure plotted against width of breakdown gap

voltage lower the higher the permittivity, but where the applied voltage is only slightly above the inception voltage in a material of medium permittivity, a cavity may grow to a point at which it ceases to discharge.

A simple analysis of the rate of discharging of a cavity in an alternating field is given by Whitehead (1951, 1954) which implies an additional four discharges per cycle whenever the field reaches a new integral multiple of the breakdown field of the cavity. However, the discharges are more numerous and increase much more rapidly with voltage than this analysis indicates, because a number of discharge sites are often located in each cavity and raising the field brings more sites into action.

With direct (as opposed to alternating) fields the frequency of discharge is determined by the resistivity of the dielectric which controls the rate at which the surface charge builds up after each discharge. With impulse voltages a single discharge or a number may be produced by a single impulse (see Salvage, 1968 for summary).

Apart from inception and extinction voltages the other quantities of interest are *discharge magnitude* and *discharge energy*. The discharge magnitude (or level) is the quantity of electricity Q (in pC) apparently lost at each discharge by the capacitance C formed by the insulation under test, it is usually less than the quantity of electricity transferred by the actual discharge in the internal cavity itself; if the voltage changes momentarily by δV when the discharge occurs, $Q = C\delta V$. The discharge energy is $\frac{1}{2}CV\delta V$, V being the applied voltage. (The factor $\frac{1}{2}$ occurs because the process is not the removal of a small charge Q from a constant capacitance C but the short-circuiting of a small element of capacitance). Methods of measuring these quantities are indicated in Chapter 6.

Discharge in oil-impregnated paper may be due to imperfect drying of the paper rather than to pre-existing voids. This is probably due to electrolytic generation of gas and is further discussed in Chapter 5. Once discharge occurs in oil-impregnated paper for whatever reason, there is a tendency for gas to be generated by decomposition of the oil, thus making the voids larger (but see Section 10.1.3).

The avoidance of all discharges at working voltage is an obvious ideal for high voltage insulation which is not always necessary nor economically justifiable. Howard (1951) found that discharges in polyethylene cable of less than 2pC were relatively harmless and there is frequent evidence that oil-impregnated bushings which have given long service are in fact discharging (Jolley, 1965).

Suggestions for meeting this difficulty range from the use of insensitive methods of detection to a scientific assessment of the balance between gas generated in discharging voids and its removal by solution in oil.

Marlow (1970) suggests that discharge limits of 10 pC for oil-impregnated bushings and 100 pC for synthetic resin bonded paper bushings at working voltage are acceptable. Constandinou (1969), working with model bushings under oil, concluded that discharge indicates the presence of trapped air but not necessarily a short prospective life if the air can dissolve in the oil.

The possibility of diagnosing deterioration of insulation from the temporal patterns and magnitudes of discharges has been investigated recently, no conclusive results seem to have been obtained.

4.3.6 Thermal breakdown

The concept of thermal breakdown is quite precise, but it is rare that breakdown can be attributed to this process unequivocally, except in laboratory experiments designed to ensure that all the relevant quantities are measurable. The concept in its simplest form is that conduction or dielectric loss in the material raises the temperature to a point at which, for one reason or another, it can no longer withstand the field. The heat generated, rising as the square of the field strength, raises the temperature of the insulation until the heat dissipated from it is equal to the heat generated; when this equilibrium temperature is high enough to cause melting, gas evolution or degradation of the material, breakdown will clearly occur. The state of affairs more commonly envisaged, however, arises from the increase of conduction or loss with temperature which occurs in nearly all materials; because of this the heat generated rises as a function of temperature, and the temperature rises as a result of the heat generated; there must then be some field strength at which the heat generation increases with temperature more rapidly than heat dissipation, there is then no equilibrium temperature and a catastrophe must ensue. Mathematical development of these arguments leads to a number of conclusions.

Obviously there are many possible circumstances of dissipation conditions, size and shape of electrodes, variation of conductivity or loss with temperature and also with field strength; calculations relating to several typical cases are summarised by Whitehead (1951). With d.c. and power frequency a.c., designers are usually able by experience to avoid materials with conductivity or loss high enough to cause thermal breakdown, but at high ambient temperatures, or with lossy materials at high frequency it may become a limiting factor. In what follows, 'conductivity' in an a.c. case is to be taken to mean $\omega\varepsilon''$, i.e. the a.c. conductivity equivalent to the total dielectric loss.

An important characteristic of thermal breakdown is that the breakdown voltage does not increase in proportion to insulation thickness. The greater the thickness the more slowly heat is removed from the central region and therefore the lower the field strength that can be supported without reaching thermal instability or a temperature of decomposition. For thin layers, if the conductivity is independent of field strength the breakdown voltage will increase only as (thickness)$^{\frac{1}{2}}$. For thick layers, the breakdown voltage cannot be increased indefinitely by increasing the thickness, an asymptotic value is reached as the thickness becomes infinite. This value can be calculated from the properties of the material and is independent of the way in which heat is removed from the outside surface so long as the electrodes are cooled by some fixed ambient temperature; an approximate value is obtained by plotting.

(thermal conductivity)/(electrical conductivity)

against temperature and measuring the area under the curve from the ambient temperature to infinite temperature (in practice the temperature at which the conductivity has increased by a factor of about three is near enough to infinity). The square of the maximum voltage that can be applied is then eight times this area (expressed in appropriate units). Examples are given by Whitehead (1951, p. 124).

The effect of increasing the ambient temperature is independent of the thickness and configuration of the insulation; provided thermal conductivity is constant and electrical conductivity is independent of field strength the thermal breakdown voltage decreases as

1/(electrical conductivity at ambient temperature)

If conductivity increases exponentially with temperature, instability is likely when the temperature rise due to self-heating is such as to increase the conductivity by two or three times.

The above relationships depend on simplifications and approximations, they also assume that heat flow is parallel to the electric field; they should not be used quantitatively without reference to fuller treatments, given by Whitehead or in the literature which he quotes.

4.4   Electric breakdown of liquids

The characteristic breakdown strength of a liquid, if it exists, should be easy to observe; liquids are readily handled and purified, they are usually transparent, it is very easy to avoid visible cavities and to ensure that boundaries with other insulants lie outside a region of high field strength. Yet the highest measured breakdown strengths lie below the highest values for solids of comparable chemical structure and density by a factor of four or more. Space charges may well occur in liquids as well as in solids but can hardly account for this discrepancy.

While the techniques of discharge-free measurements were being developed for solids, experimenters working on liquids developed very elaborate purifying, drying and filtering techniques, and studied the effects of electrode materials and preparation (including oxidation), the time lags associated with breakdown, and the effect of hydrostatic pressure and other variables on breakdown of organic liquids and of liquefied helium, nitrogen, argon, oxygen etc. Breakdown stresses around 0·1–0·2 MV per mm were generally found. These, and earlier results on normal insulating liquids, are reviewed by Sharbaugh and Watson (1962) and by Lewis (1959). Recent results have been reported by Jefferies and Mathes (1970) (nitrogen and hydrogen), Meats (1972) (helium), Spitzer (1970) (transformer oil) and Boone and Vermeet (1970) (fluorocarbons); further references will be found in these papers (see also Section 10.8). Generally, decrease of hydrostatic pressure lowers the breakdown strength, increase of pressure often raises it. (For this reason sealed oil-impregnated transformers and capacitors should be so designed that the pressure within them is never less than about 1 atmosphere.) Various writers have associated this with the presence of dissolved gas, but the evidence is conflicting. Dissolved oxygen may act in the same way as oxidation of the electrodes, to increase the breakdown strength.

The most common type of mechanism suggested for liquid breakdown has been collision ionisation leading to avalanches, analogous to gas breakdown; but more recently the possibility has been studied that the formation of a vapour or gas bubble initiates the breakdown, rather than follows it.

The idea that the electric field of itself causes vapourisation or encourages gas bubbles to form is not tenable. Many textbooks of electricity are confusing on the subject of the forces produced by the field at a gas–liquid interface; but Garton and Krasucki (1964) established that an insulating bubble in an insulating liquid is subjected to an *increased* pressure by the field, with a tendency to elongation along the direction of the field. They calculated the change of shape in terms of prolate spheroids and concluded that the elongation is not very great unless the ratio of permittivity of liquid to gas is very high. If the bubble becomes conducting however the field tends to enlarge it, and it may elongate indefinitely. A bubble with a discharge in the vapour or gas within it can therefore both grow and elongate. A bubble initiated on an electrode would, if the field were sufficiently high, have its inner surface charged by emission or discharge from the electrode and would tend to grow along the field. Krasucki (1962) found that in a very viscous liquid (hexachlordiphenyl) vapour bubbles can be observed to grow from a pointed electrode, and that if heating were avoided by using short pulses a tree-like breakdown pattern, like that in a solid, developed. In further experiments on the same material Krasucki (1966) found that the rate of growth of a bubble is consistent with the assumption that all the energy of the current flowing into it is used in forming vapour. Using spherical electrodes he found strong evidence that a change of impulse breakdown strength from 0·5 to 0·1 MV/mm was attributable to a change of viscosity by several orders of magnitude (which in this liquid occurs over a small temperature range), and that if a constant stress of 0·12 MV/mm was applied the time to breakdown was proportional to viscosity from 10s at 10^7P to 10^{-5}s at 10P. This behaviour was to be expected from calculation if breakdown occurred when the bubble reached a certain size. For light mobile liquids he concluded on the same basis that breakdown would be determined by surface tension rather than viscosity and that the effects of temperature and pressure, also the magnitude of the formative time lag, are consistent with this. It thus seems that, with some liquids at least, breakdown could be explained by the initiation of a bubble of the vapour of the liquid at a site at which electrons are injected into it, followed by purely electromechanical consequences.

The well-known increase in the impulse strength of impregnated paper with viscosity of the impregnant and the impermeability of the paper may be related to these observations.

In an extremely divergent field near a point of radius 0·1 to 100 μm bubbles are 'sprayed' from the point at voltages approaching those at which breakdown occurs (Krasucki, 1962; Singh *et al.*, 1972). The existence of bubbles which do not initiate a breakdown is presumably explicable by their moving out of the region of strongest field before they have had time to expand catastrophically.

It must not be forgotten that in practical circumstances breakdown is frequently associated with foreign particles, especially cellulose fibres. It is still possible, even with refined techniques for removing visible particles, that submicroscopic ones play a part in breakdown. Krasucki (1972) has found that washing a test cell in filtered hexane does not clean the cell of particles of a few μm diameter, the application of ultrasonic agitation results in quantities of fresh particles being detached from the cell surfaces, and the repetition of washing and ultrasonic treatment merely reduces the numbers of such particles by a factor of two or three each time.

It is also clear that measured breakdown strengths of liquids diminish as the volume of liquid under stress is increased, and as the period of stressing is lengthened.

Both these effects are referred to again in Section 10.1.5.

4.5 Tracking and tree-burning

The processes by which organic insulation is sometimes rendered useless by the formation of carbonaceous layers or lines on its surface are somewhat varied, but are basically the charring of the material by some form of discharge.

One such process is simply the presence of an arc, carrying at least one or two amperes, burning close to the surface of a material such as a phenolic resin-bonded board; it immediately produces a patch of pyrolised resin which is conducting, and if the arc was passing between two metal electrodes attached to the board a good conducting path will remain between them after the arc is extinguished. This type of tracking is the one to which arc-chutes must be resistant.

A second type of process occurs when a relatively small current at high voltage, in the form of spark discharges, passes between two electrodes attached to the surface of, say, a phenolformaldehyde board. It is probable that the board, if clean and dry, will not suffer for a considerable time, but if it is wetted by spray or condensation one or more thin wavy tracks of charred resin will form, which, if the discharge and wetting continue, may form a complex tree pattern ultimately bridging the electrodes (Fig. 30) (Plate).

A third type of process helps in understanding the preceding one: a solution in water of any salt is spread in the form of a continuous film (with the help of a wetting agent if necessary) between two pieces of metal resting on a phenolformaldehyde paper board and a supply of 200–300 V (with current limiting resistors of about 1 kΩ) applied to the metal pieces. The film is evaporated by the heating of the current, the thinnest region gets heated the most and rapidly thins down to a narrow neck which breaks with a tiny bright discharge like a miniature arc, but it need only be carrying a few milliamperes. A small spot of charred material is left where the discharge occurred. If the liquid then flows back or is replenished the process repeats, and with repetition the spot becomes extended into lines more or less along the direction of the field. If the liquid is steadily replenished at about the same rate as it evaporates pools remain near the electrodes separated by a boundary at which the film is being broken and reformed continuously, this region presents a relatively dry appearance with scintillation each time the circuit is broken and is called the 'dry band'. Ultimately the carbon tracks starting in the dry band will lengthen, broadening the dry band until one of the tracks bridges the electrodes and short circuits them. If the limiting resistor has been omitted an arc limited only by the supply impedance will follow, the fine structure will be obliterated and the result will be indistinguishable from that due to high voltage flashover followed by an arc. Dry bands can be formed whether the underlying material tracks or not, and can be observed on porcelain insulators under heavy deposition.

A fourth type which has often been observed takes place in oil containing droplets of water. These are drawn into the strong field on the surface of a bushing and form a layer on the surface which behaves in the manner described above. A decorative example of this is shown in Fig. 31 (Plate).

Tracking of the first type is normally the result of a particular incident such as a flashover, or of a wrong choice of material exposed to arcing conditions, or close to a hot component.

Tracking of the second, third and fourth types may take years to complete in service; in the case of the second and third types the aqueous conducting path may only occur when humid conditions or fog cause deposition on the surface. Dust on an insulator surface, especially if it contains electrolytic substances, favours tracking by spreading moisture in a film; where the dust is itself conducting, e.g. carbon dust from motor brushes, it may act somewhat analogously to a water film even when dry. A clean surface on the other hand, especially a glossy or water repellent surface, keeps the moisture in discrete drops which do not allow current to flow.

Obviously the inorganic materials cannot form carbon tracks unless covered with a layer of oily dirt, even then the resulting track will be disrupted by any substantial short circuit current leaving a somewhat rough and pitted surface which is easily fouled again. Among the organic materials there are wide differences in susceptibilty to tracking. Organics which melt or volatilise without charring are unlikely to track unless covered with dirt, those which produce large quantities of gas or vapour when they char, or form a powdery non-conducting deposit are relatively resistant to tracking, while those which form a firm solid char are among the worst. There is a good deal of evidence that aromatic rings in a polymer molecule favour carbonisation. Table 1 summarises roughly the relative susceptibilities of a number of common types of material as indicated by the British Standard test corresponding with the third type of tracking described (Section 6.3.8). Parr and Scarisbrick (1965) have put forward considerations of relative bond energies affecting the formation of carbon and of the possible action of hydrated alumina filler as providing steam to oxidise carbon or blow it away.

Materials which are prone to tracking by moisture deposition can be given some protection by a layer of a less susceptible varnish, glyptals have often been used. Surprisingly an inorganic pigment does not necessarily improve, and may worsen, the tracking properties of a varnish.

The tendency to tracking by moisture appears to be the main obstacle to the outdoor use of rigid organic materials. There are indeed others such as deterioration in sunlight and endurance to extremes of temperature, but tracking seems to be the most serious difficulty. Materials which are extremely resistant to tracking when new may succumb to it after a period of dry band erosion.

Field tests have been carried out which indicate that epoxides of the type known as cycloaliphatics (see Section 9.5.1) anhydride-cured and with hydrated alumina filler can give satisfactory life as overhead line insulators provided the electric stress is not too high (Billings and Humphreys, 1968; Dey, Drinkwater and Proud, 1968; Stannett *et al.*, 1969). [Some authors, however (for example Palumbo *et al.*, 1967), find the advantages of cycloaliphatic resins over bisphenol ones less clear-cut.] Failure takes place by erosion rather than track formation, at a rate controlled by weathering more than by the presence of the electric stress. The additional length to give a low stress could be obtained by making the tower cross-arms into the insulator or by increasing the suspension insulator length. The relative lightness of epoxy insulator sheds and the possibility of having a strong epoxy glass core, makes longer suspension insulators mechanically possible. It is possible

Table 1   Comparative tracking indices of some insulating materials measured according to BS 3781

(Based on data provided provided by the Electrical Research Association and by GEC Ltd.)

Resin	Filler or reinforcement	CTI
Phenolic	Paper, cotton fabric, nylon fabric, wood, rubber	80–120
Polycarbonate	—	
Phenolic	Asbestos paper or fabric, glass fabric	
Polyphenylene oxide	—	120–220
Phenolic-epoxide	Mica	
Epoxide*	Silica and alumina or glass fabric or fibre	
Butyl rubber–p.v.c. copolymer	—	
Alkyd	Mineral or glass fabric	
Polystyrene	—	
Various polyesters	Glass fabric	220–300
Silicone resin	Glass fabric	
Polyimide film	—	
Melamine	Glass or asbestos fibre	
Urea formaldehyde	Wood flour	
Diallyl phthalate	Glass fibre	
Nylon	Glass fibre	
Melamine	Glass fibre	300–500
Alkyd	Mineral	
Butyl rubber	—	
Rigid p.v.c.	—	
Urea formaldehyde	—	
Epoxide	Asbestos	500–700
Silicone rubber	—	
Acetal resin	—	
Polypropylene	—	
Polyester	Glass fibre	
Polymethyl methacrylate	—	greater than 700
Polyethylene	Unfilled or carbon filled	
Polytetrafluorethylene	—	

* This table does not include alicyclic epoxide resins with hydrated alumina filler which have a very high CTI

that in highly polluted areas long epoxy-glass insulators may be more reliable than porcelain due to a lesser tendency to flashover when dirty.

Plastics which are highly resistant to tracking and weathering, such as polyethylene, polytetrafluorethylene and silicone rubber, are not stable enough mechanically to be used for insulators, but the possibilities of coating epoxies with silicone rubber, without separation at the interface occurring, are being tried (Bradwell *et al.*, 1970).

Chapter 5

Moisture in insulation

5.1 Effects of moisture

The effects of moisture on the electrical properties of insulation are so well known that they are often accepted as axiomatic. The commonest are worsening of all electrical properties and a general tendency to worsening of mechanical properties; in cellulose there are expansion, increase of flexibility, and increase of strength as well as deterioration electrically.

In general, water has little effect on organic materials which it does not wet and which have no affinity with it; the *lyophobe* materials, which include the hydrocarbons, polytetrafluorethylene, silicone rubbers and non-dipolar materials generally, unless they are impure. A very pure hydrocarbon oil exposed to humid air shows a loss angle less than 0·0002 at all frequencies up to 10 MHz, but if it contains a small proportion of an ester a variety of conduction effects appear on exposure to humid air (Dunkley and Sillars, 1953). Paraffin wax containing droplets of water shows only the Maxwell–Wagner loss peak associated with spherical particles, so long as the wax is continuous, but if intercrystalline fissures containing relatively small volumes of water are present a large loss angle appears at all frequencies up to the Maxwell Wagner limit (Sillars, 1937; see also Section 4.2.3.1). It is not surprising therefore that the quantity of water contained in a solid is not a reliable guide to the degree of deterioration of electrical properties produced in different materials, or even in similar materials under different conditions.

The immunity of intact lyophobe materials to the effects of water may need some qualification in the case of breakdown. Polyethylene-insulated cable can be used under water without apparent effect, and cables with cooling water flowing down the hollow stranded conductor (and therefore in contact with the insulation) are in regular use, but there have been instances of polyethylene and other hydrocarbon polymers breaking down at quite low a.c. stresses under water, due to the growth of trees similar in appearance to those formed at high voltage under dry conditions (Miyashita, 1971). No explanation of this has been found, but there seems to be good evidence that it only occurs with certain grades of polymer.

Organic materials which have affinity for water, such as those containing hydroxyl, carboxyl and amine groups, which are wetted by water (or, if liquids, are miscible with it) are called *lyophil* materials, and are not intrinsically suitable as electrical insulation. Cellulose materials are, of course, in this category and are extremely important and widely used for their cheapness, flexibility and ease of application, but they cannot be used in any situation involving appreciable electric stress without protection from the atmosphere by impregnation with resins or oil or by other means. Despite its low cost, cellulose paper is losing ground to materials which need no protection from moisture.

Intermediate substances such as those containing ester, ether, ketone and amide groups, including a great variety of natural and synthetic resins, are noticeably affected by moisture and may or may not need protection from the atmosphere according to the use to which they are put.

Impregnation of fibrous materials by resins is a satisfactory way of preventing the major effects of moisture, but does not entirely avoid them. Polymeric materials, unlike metals, have a measurable permeability for moisture which varies greatly from one material to another; so that prolonged exposure to damp conditions, or weather, or contact with liquid water, will have a deleterious effect on the best impregnated paper or asbestos laminates. Moreover, a degree of impregnation sufficient to confer reasonable mechanical rigidity on a laminate is not sufficient to protect it from prolonged humidity as encountered in tropical conditions. Coating with varnish merely delays the ingress (and egress) of moisture, smoothing out fluctuations without altering the effects of humidity averaged over days or weeks. The permeability of resins and polymers to moisture is difficult to measure accurately and depends on both physical and chemical factors, but broadly materials which absorb the least water also transmit the least, the best being hydrocarbons; epoxies, polyesters, polyvinyl chloride and phenol formaldehydes are somewhat worse, and cellulose esters are worse still. Cellulose itself can be regarded as having unlimited absorption. Some comparative figures for moisture absorption and permeability are given by Clark (1962).

Impregnation of cellulose materials with hydrocarbon oil is permanently effective only if the oil itself is isolated from atmospheric moisture.

The effects of moisture are worsened by traces of salts, acids, alkalis, alcohols or any substance which ionises in solution in water; these must therefore be excluded from any material intended as insulation; thus paper during manufacture must be washed free of all salts and alkali, phenolic resins should be fully cured, insulation exposed to smoke or salt-laden atmosphere should be cleaned periodically, and so forth.

The reasons for the effects of water can be guessed at in general terms but have never been elucidated in detail. The moisture may in some cases be homogeneously disposed (i.e. in solution) throughout the solid or liquid; but in many materials it is more likely to be adsorbed on external and internal surfaces and absorbed by cellulose fibres. As example of the former type, Reddish (1950) has pointed out that the loss tangent against frequency curves for damp polyethylene terephthalate are entirely similar to, but higher than, the corresponding curves for the dry polyester (Fig. 17). The loss in the dry polyester is attributed to the rotation of the unreacted hydroxyl groups at the end of each ester chain in an amorphous region, the additional loss is attributed to hydroxyl groups of the added water molecules in similar environment. In most materials, however, the effect on loss tangent is spread over a wide range of frequencies; in the case of water in cellulose fibres, for instance, one may speculate about distributed relaxation times of dipoles, or some form of restricted conduction limited to single fibres or parts of fibres, or co-operative hydrogen bond effects, all with plausibility since there is no evidence confirming or excluding any of them, and purely electrical measurements cannot provide this.

The effect of moisture in reducing breakdown strength cannot be explained without a better understanding of breakdown processes.

It is intriguing to note that Mathes (1969) has observed that in liquid hydrogen (20 K) the presence of moisture has, if anything, a beneficial effect on loss tangent and breakdown strength of paper impregnated with this fluid.

Oil-impregnated paper which has not been fully dried before impregnation shows a low discharge inception voltage even though initially there may be no bubbles or cavities present. This is due to the formation of bubbles of gas by electrolysis of the water absorbed in the fibres, often at a point of contact between a fibre and one of the conductors to which the paper has been applied (Krasucki, Church and Garton, 1960, Kogan, 1963). If the rate of generation of gas exceeds its rate of solution in the oil, a bubble will grow until it is large enough for discharge to occur at the prevailing electric stress, after which progressive breakdown may develop.

The importance of eliminating moisture from paper is most crucial in oil-impregnated high voltage insulation, namely in bushings, transformers, cables and high voltage power capacitors. It is natural to want to know the quantitative effects of moisture upon both oil and paper. There is considerable indefiniteness about the solubility of water in oil and the quantitative effects on conductivity, loss tangent or electric strength; all of these depend on the content of aromatics, impurities and dust particles in the oil. Generally the solubility is of the order of tens of parts per million at 20°C, rising by a factor of ten or so at 100°C. There is evidence that moisture in solution does not affect the electric strength of oil in the complete absence of solid particles, but there is no doubt that oil containing a normal quota of suspended particles after the usual processing techniques may lose 50% or more of its electric strength if the water content increases from 10 ppm to 50 ppm (Binggeli *et al.*, 1966). These questions are considered further in Section 10.1.4.

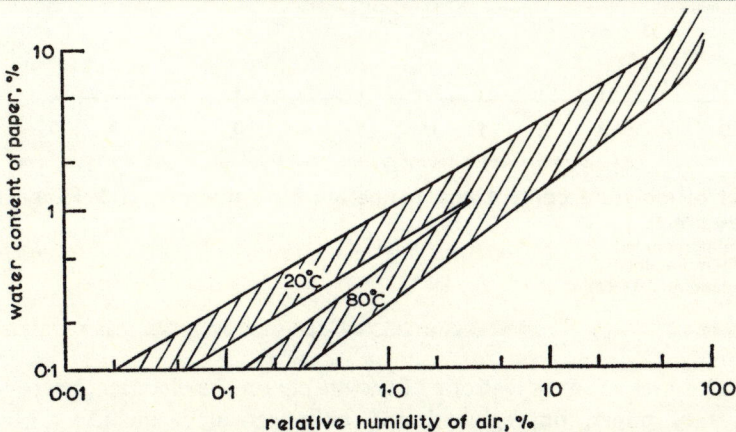

Fig. 32 Approximate range of moisture contents of typical papers plotted against relative humidity of air in equilibrium with them

Note that for the same relative humidity the content of water in air at 80°C is about 20 times that at 20°C. Based on data by Piper (1946) and others. (See Fig. 47 also)

D

The equilibrium vapour pressures of cellulose materials with a wide range of water contents have been studied by Piper (1946). The approximate spread of the curves of relative humidity of air vs water content of paper in equilibrium with it is given by Fig. 32. Typical curves of water content against vapour pressure are shown in Fig. 47. In practice there is hysteresis in approaching the equilibrium, so that the apparent equilibrium content will be lower when the humidity has been increased from a low value, and vice versa.

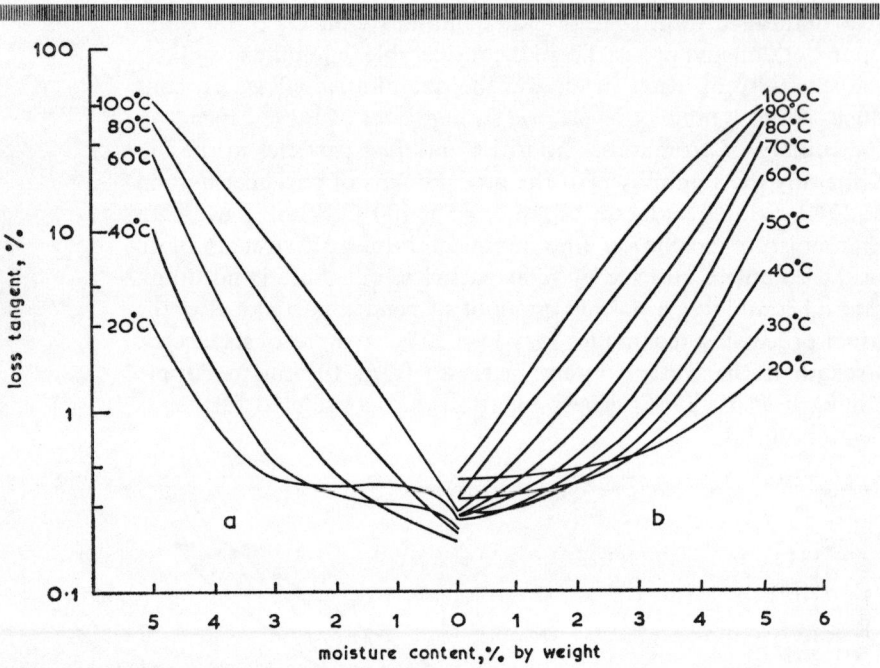

Fig. 33 Effect of moisture content and temperature on the loss at 50 Hz of a dense paper
(a) Unimpregnated
(b) Oil-impregnated
(Constandinou, 1965b)

A series of investigations of the effects on the electrical properties of paper, impregnated and unimpregnated, including loss tangent, dispersion (change of capacitance with charging time, Section 6.3.5), insulation resistance and permittivity has been summarised by Constandinou (1965b); more detailed results are available in the reports quoted in his paper. Figs. 33 and 34 are

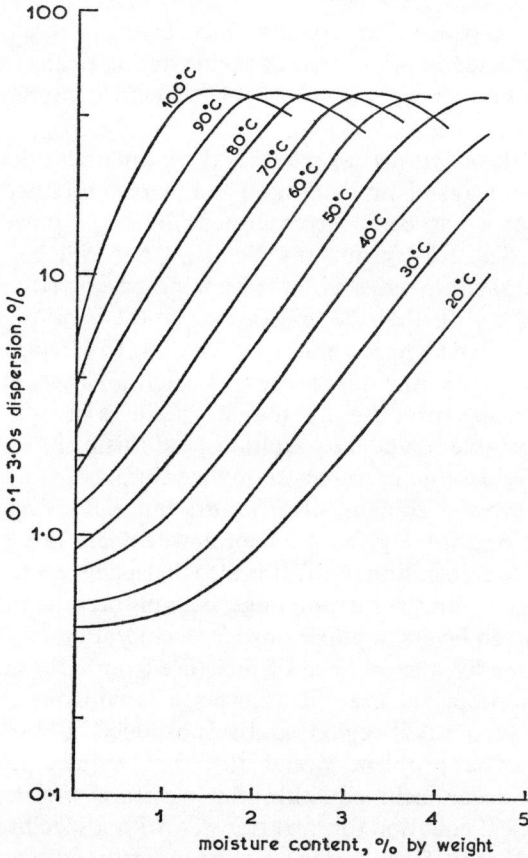

Fig. 34 Effect of moisture content and temperature on the 0·1–3·0 s dispersion of an oil-impregnated paper (Constandinou, 1956*b*)

reproduced here as examples. Electrical measurements do not usually give very reliable indications of moisture content at room temperature, because the rate of variation of electrical properties with moisture is low in the region of greatest interest, i.e. below 1%. The effects of moisture on a.c. and impulse electric strength of impregnated paper have been studied (e.g. Binggeli *et al.*, 1966) but are probably of less importance than the effect on discharge inception voltage.

Standard methods of measuring water content are indicated in Sections 6.4 and 6.6.

5.2 **Removal of moisture from high voltage insulation**

It is difficult and expensive to dry out thick layers of paper once it has been immersed in oil, so that adequate drying to ensure that discharge does not occur must be completed before impregnation.

This takes many days for the layers several centimetres thick required at the higher transmission voltages, and there is no direct way of checking that it has been accomplished. Since the innermost layers are the last to dry out any electrical test will be a measurement on dry paper in series with the remaining damp paper and this accentuates still further the insensitivity to the condition of the inner layers. At first sight it would seem possible to calculate the moisture content of the inner layer from the conditions used in the drying cycle, starting from the differential equations of diffusion and applying suitable boundary conditions, but this turns out to be difficult. Firstly the rate of diffusion of water in paper is not simply proportional to the gradient of concentration and a function of temperature but, for a given gradient and temperature, is lower the lower the concentration itself. It is also dependent on the total gas (or air) pressure in the surroundings, perhaps because the presence of gas between layers of paper impedes evaporation from one to the next. Secondly, the removal of moisture from a region removes also the appropriate heat of desorption (analogous to heat of evaporation) from that region, and this produces a lower temperature there. The problem would therefore require the solution of a non-linear field equation for moisture transfer coupled to a linear field equation for heat transfer, with a specified rate of supply of heat at surfaces as one set of boundary conditions and a rate of removal of water dependent on the resultant surface temperature as another.

Lampe (1969) has reported calculations of moisture concentration as a function of insulation thickness and distance from surface, temperature, and drying time, assuming that the temperature is maintained uniform throughout the insulation, and that the rate of diffusion rather than the rate of surface evaporation is the limiting factor in moisture removal.

The practice of drying insulation is based largely on experience coupled with observation of the total rate at which water is removed (by condensing it) and of an electrical property of the insulation being dried. Essential requirements are the supply of heat and a final drying in vacuum.

5.2.1   Transformer drying

The initial stages are usually carried out in circulating hot air or gas, to convey the heat needed for the evaporation of the major part of the water; the later stages require vacuum pumping equipment able to remove adequate volumes of water vapour at a low enough pressure to ensure drying to a moisture content below 1 % in a reasonable time. During vacuum drying heat is supplied by circulating currents in the windings, or by radiation, or by periodic readmission and circulation of hot air or gas (see Dickson, 1959, for a summary of various processes in use then).

There are various opinions about the lowest pressure required, Binggeli *et al.* (1966) suggesting a final value of 0·1 torr (about $13\,N/m^2$) as satisfactory, Kettler and Lange (1956) on the other hand regarding 0·001 torr as desirable. Generally a moisture content just less than 0·5 % is aimed at, though figures down to 0·2 % are often quoted; transformers will give satisfactory service at higher moisture contents but it is inadvisable to rely on this. It seems to be accepted that the time required is 200 or more hours for the highest voltage transformers. Beer *et al.* (1966) have carried out drying tests on models of transformer insulation up to 36 mm thick, plotting loss tangent and pressure in the tank during drying at various temperatures. Kusay (1971) describes a vacuum plant suitable for drying large transformers to a high standard.

Impregnation after drying is by admitting oil while maintaining the vacuum.

Alternatives to the simple vacuum drying process, with periodic reheating by hot gas, are the oil spray and the vapour phase drying processes. In the former hot dry transformer oil is sprayed over the windings in the vacuum chamber; this raises the temperature rapidly and maintains it, but the presence of oil in the paper impedes removal of moisture. In vapour phase drying the vapour of a pure light petroleum liquid is produced in the vacuum chamber and condenses on the windings: the condensate runs back and is re-evaporated. The windings are heated to the condensation temperature of the vapour, usually 130–140°C, water is removed from the paper by a process which is more rapid than diffusion through oil-soaked paper, and is evacuated by the pumps. This process is considered to be quicker than simple vacuum drying, this may be because the former is carried out at 130–140°C apparently without undue degradation of the paper, whereas the latter is limited to about 105°C. The vapour phase method is very advantageous for redrying transformers which have become wet after impregnation, since the transformer oil is washed out by the light fluid and does not impede drying. None of the methods has established an overall preference over the others in normal manufacture; see the summary by Hetherington and Keil (1966).

Drying of transformers which have become damp can be carried out very slowly by circulating the oil through a vacuum drying plant, or more economically through a bed of material known as 'molecular sieve', a specially prepared calcium or potassium aluminium silicate of defined crystalline form having a regular system of pores in the crystal structure of the right size to accept water molecules. In either case the rate of removal of moisture is limited by the rate of transfer from the paper to the oil, which is very slow, requiring many weeks to produce a substantial change of moisture content of the paper (see Franklin, 1965). This method is used for drying while on load.

5.2.2 Cable drying

Cables for voltages up to about 33 kV to be impregnated as 'solid' cables are dried in a straightforward way by heating the paper-lapped unsheathed cable in an oven at 110°C, then vacuum drying at a similar temperature and finally admitting the impregnating compound. When impregnation is complete the cable is cooled and the sheath applied to the now solid insulation. Electrical tests (loss tangent or insulation resistance) are used to monitor the drying process.

Drying of high voltage oil-filled paper cables, on the other hand, is more critical than transformer drying, for the stress at working voltage is considerably higher and therefore discharge is more difficult to avoid, also dielectric loss is important both as a waste of power and because it reduces the current rating of the cable. The older methods of drying through the oil duct at the centre of the conductor after sheathing have been largely superseded by the use of equipment which vacuum dries the insulation after taping and then applies the sheath under vacuum or under oil without exposure to air. Moisture contents down to 0·1% are usually considered desirable.

Moisture has a noticeable effect on the dimensions of paper which may assume importance where thick layers of paper are wound; the redistribution of stress on drying may have to be allowed for in deciding on the winding tension used. Typically, drying paper from a 5% moisture content reduces the diameter of the fibres by 2–3% and their lengths by about one-tenth of this.

Chapter 6

Test methods

Descriptions of test methods are frequently misleading unless full and precise; this chapter therefore gives only the explanations necessary to show the general significance of the most frequently quoted properties, and refers to standards or other literature for details. It includes only the commoner types of tests applicable to large classes of finished materials; tests used only for individual materials or small classes of materials, or for materials before or during processing, must be sought under the names of these materials or classes in the index of particular national standards; a procedure which may involve trial and error in some systems, notably the British.

The national standards of the USA and Germany are quoted as well as British and international standards, where it seems useful to do so, but this is not a complete guide to standards relating to the subjects quoted nor a comprehensive concordance of them. Methods of test may differ in detail according to the material tested, and in the British system many of these differences are not presented by indexing or cross-referencing.

Mention of the specifications of different nations in describing any particular type of test does not imply that these specifications agree in detail unless agreement is specifically stated.

For the simpler types of well established arbitrary test (those tests which measure a representative aspect of practical behaviour rather than a scientifically significant quantity) only standards are quoted. For tests of quantities which can be defined apart from particular measuring techniques, e.g. dielectric loss, and where development of methods is therefore possible without losing comparability, background literature is also quoted.

British standards are identified by the prefix BS; standards of the American Society for Testing and Materials have simply the prefix C, D or E as in the ASTM designation; those of the International Organisation for Standards have the prefix R as in the ISO system; those of the International Electrotechnical Commission have the prefix IEC; those of the Deutscher Normenausschuss have the prefix DIN. The recommendations of the US Institute of Electrical & Electronics Engineers and of the German Verband Deutscher Elektrotechniker are prefixed by IEEE and VDE respectively.

It is important that the determination and presentation of data on the properties of plastics takes account of all the relevant variables and conditions. BS 4618: 'Recommendations for the presentation of plastics design data' gives guidance on this in relation to several of the mechanical and electrical properties of interest here.

6.1 Mechanical properties

We have seen in Chapter 2 that the mechanical properties of polymers are affected by the time-scale of the test as well as by temperature and often by humidity. It is generally important to specify ambient conditions preceding and during the test and to control the rate of extension of, or the rate of applying load to, the specimens. Testing machines which enable both load and extension to be recorded, preferably plotted automatically, are necessary.

6.1.1 Tensile properties (short time)

Definitions of terms, conditions of testing, equipment, dimensions of specimens, also procedures for determining tensile strength at yield or break, percentage elongation and elastic modulus are given in relation to particular types of plastic material by BS 2782–301A to L, and BS 2782–302A; corresponding definitions and procedures are given in more general terms in D 638. Elastic modulus as defined by BS 2782 is measured at 0·2% strain; as defined by D 638 it is the 'initial slope' of the stress/strain curve. R 527, DIN 51221 (test machines) and DIN 53455 (plastics) also deal generally with tests of tensile properties. Other relevant standards are: D 651 (moulded materials); D 882, D 1708, D 1923, BS 2782–302C and R 1184 (thin films); D 902 (coated glass fabrics); D 2105, D 348, D 349 (tubes and rods); BS 903–A2 and R 27 (rubbers). See also D 1822 under Section 6.1.3 (tensile impact), and note that beam flexure properties (Section 6.1.2) are related to tensile properties.

6.1.2   Flexure properties and cross-breaking strength

General methods for measuring elastic modulus, yield and cross-breaking strength by deflection of a beam supported near its ends and loaded at its centre are given by BS 2782–304A to E, R 178, D 790, DIN 51277 and DIN 53452. BS 2572 deals with phenolic laminates, BS 3497 Appendix K with asbestos board, BS 1598 and C 674 with ceramics. D 747 prescribes a rotating cantilever method for elastic modulus of non-rigid plastics and D 797 a method for elastomers.

6.1.3   Impact properties

General recommendations for methods of measuring and presenting impact data are given in BS 4618, section 1.2. The commonest types of impact test are those in which a notched cantilever bar (Izod) or notched beam supported at both ends (Charpy) is struck by a calibrated pendulum, and the net energy required to break the specimen is measured. (The width of the specimen, perpendicular to the direction of striking, depends on the dimensions of material available, and in this book the values quoted follow the practice of stating energy per unit width, Joules per metre corresponding with ft/lb per in; BS 4618, however, recommends reduction to energy per unit area fractured, J/m^2.)

Izod tests for rigid plastic materials are described in BS 2782–306A and D 256A; Charpy tests in BS 2782–306D, BS 2782–306E (similar to R 179) and D 256B. Pendulum tests generally are dealt with in DIN 51222 and DIN 53453.

Impact testing by dropping a weight on the centre of a disc-shaped rigid specimen supported round its perimeter is described in BS 2782–306B and C; analogous tests of a flexible film held in a circular support and struck by a falling dart are in BS 2782–306F and D 1709.

A test for the energy required to break a specimen by tensile impact is described in D 1822.

6.1.4   Compression properties

Usually compression failure is of more interest than compression modulus. Various shapes of specimens are prescribed, usually with the dimension in the direction of compression equal to, or somewhat greater than the dimension perpendicular to it.

BS 2782–303A to C, D 695, R 604 and DIN 53454 deal with plastic materials; the BS does not specify a rate of load application, but the others do. BS 1598 and C 528 deal with ceramics. D 621

D*

describes tests for permanent deformation of plastics after a prolonged period of loading ('cold flow').

6.1.5 Tensile creep and stress-relaxation

The long-time extension under load, or relaxation of stress at fixed strain, of polymeric materials is very sensitive to temperature and humidity and to fluctuations in these. Precautions and methods of presentation of data are described in BS 4618–Pt. 1, section 1.1, D 674, D 2927 and R 899. BS 903–A15 deals with rubbers.

General background of long-period measurements on plastics is given by Riddell and O'Toole (1968) and Ogorkiewicz (1970*a* and 1970*b*).

6.1.6 Hardness and indentation

There are several recognised tests for hardness, each with variants; the types mostly used for plastics are known as Rockwell and Shore. Rockwell tests measure the permanent indentation when the load on the indenting ball is increased from a small to a large value and (usually) reduced to the small value again; they are generally applied to plastics other than very soft ones and rubbers. The Shore tests measure the indentation with the load still applied to the indenter, which for a resilient and isotropic rubber, is directly related to its elastic modulus. Thus while Rockwell (like Brinell) hardness measures the resistance to permanent deformation by a local force, Shore hardness may be regarded as a convenient measure of elasticity.

Rockwell tests are designated R, L, M, E, K . . . in order of increasing severity, the severity being increased by increasing the loading and decreasing the diameter of the steel ball which (for plastics) forms the indenter; for example R uses a 12·7mm ball and a load of 60kgf, M a 6·35mm ball and a 100kgf load, E a 3·17mm ball and a 100kgf load. D 785 gives full details for plastics, BS 891 and R 80 describe the test but not the procedures for plastic materials; D 530 describes a Rockwell test for hard rubbers. DIN 53456 prescribes a broadly similar type of test for plastics using a 5mm ball.

Shore, or durometer, tests use an indenter in the form of a truncated cone with an end of diameter 0·79mm (Shore A) or a round-ended cone of radius 0·1mm (Shore D), which are pushed into the rubber by springs of specified calibration (differing between A and D). The hardness is 100 when the spring is compressed by the full available travel of the indenter without penetrating the specimen, and zero if the indenter penetrates the specimen fully with no compression of the spring. The intermediate values clearly depend on the relationship between the properties of the sample and the force-compression calibration of the spring. D 2240, R 868 and DIN 53505 describe the instrument and procedures.

BS 903–A26, R 48, D 1415 and DIN 53519 define the International Hardness Degree scale in such a way that the IRHD approaches zero for zero elastic modulus, and 100 for infinite modulus, and is approximately equal to the Shore durometer reading; they also give a curve relating IRHD to elastic modulus. The method of measurement used in these specifications is analogous to that of the Rockwell test; the indentation produced by adding a 500 gm (approx.) load on a 2·5 mm ball is converted to IRHD using a table of values.

6.1.7   Tear resistance

These are designed to assess the liability of a thin sheet to tear when tension is applied in such a way as to concentrate the stress at one point on the edge of the specimen. BS 2782–308A, D 1004, BS 903–A3 (agreeing with R 34) use a special crescent-shaped test sheet, with a nick or sharp corner at the centre of the concave side of the crescent, and measure the pull applied lengthwise required to initiate a tear. BS 2782–308B and D 1922 are methods to measure the energy absorbed from a pendulum in tearing by applying a force at the edge of a sheet perpendicular to the plane of the sheet (as in tearing between the hands) and hence find the average tearing force. D 1938 prescribes the plotting of a force/time relationship during tearing.

Tests for papers and woven fabrics are given in BS 2689, BS 4253, D 827, D 2261 and D 2262.

6.1.8   Abrasion resistance

There are several tests using various devices for rubbing the sample against a rotating surface faced with abrasive or fed with loose abrasives, the loads and speeds being defined, and the diminution of weight or volume for a given amount of rubbing being measured. Examples are BS 903–A9, D 1630 and DIN 53516 (rubbers), and D 1242 (plastics).

6.2 Thermal limitations

It is necessary to distinguish between two kinds of limit, those which are immediately evident on reaching the temperature concerned, such as becoming to soft or too brittle (these are dealt with in Sections 6.2.1–6.2.4) and the high temperature limits imposed by the possibility that the material will fail in its function due to thermal degradation before the end of the expected life of the equipment concerned. The principles underlying laboratory tests to determine the probable life of a material under given conditions have been explained in Section 3.1; the guides, recommendations, rules and specifications which have arisen from applying these principles are outlined in Section 6.2.5 and its subsections.

6.2.1 Temperature of deformation or softening

Most polymers become incapable of supporting useful loads at some temperature which is below any definable melting point; a number of arbitrary tests have therefore been evolved to compare temperatures at which materials cease to be usable on this account. In all of these the rate of heating is an important feature of the test.

6.2.1.1 Heat deflection (or distortion) temperature

A beam supported at both ends and loaded at the centre to a specified maximum fibre stress, then heated at a specified rate until a specified deflection occurs, is used in BS 2782–106G and H which agrees with R75 and is very similar to D 648. All three of these methods specify similar dimensions and procedures, with two allowable standard maximum fibre stresses (calculated from load and dimensions):

$$18 \cdot 5\,\text{kgf/cm}^2 \ (264 \text{ psi}) \quad \text{and} \quad 4 \cdot 6\,\text{kgf/cm}^2 \ (66 \text{ psi}).$$

The standard deflection at which the temperature is taken differs a little between R 75 and D 648.

6.2.1.2 Cantilever deformation tests

A horizontal cantilever with a specified load at the free end is used in BS 2782–102A and B; the deflection which occurs at a specified temperature after a specified time is found. BS 2782–102C specifies a loaded horizontal cantilever also, this is heated at a specified rate until a 30° deflection occurs.

The Martens test, DIN 53462 and DIN 53458, also VDE 0302, is similar in principle but the cantilever is held vertically with a metal rod held horizontally by a clamp at its free end; a bending moment is applied by a weight hanging on the rod. The maximum fibre stress in this test is 50 kgf/cm². The specimen is heated as specified until a specified deflection occurs.

6.2.1.3   Vicat softening point

This is the temperature at which the flat end of a vertical rod of 1 mm² section, resting on the horizontal surface of the material under test, the rod being loaded at its top end with 1 kg, sinks 1 mm below the original surface; it is detailed in BS 2782–102D, agreeing with R 306, also in BS 2782–102F and G and in D 1525. VDE 0302 is very similar but with a load of 5 kg.

6.2.1.4   Ring and ball softening point

This is the temperature at which a disc of the material under test, supported round its perimeter by a metal ring, allows a steel ball resting on the upper surface of the disc to sink through the ring. It is described in BS 2782–103A, BS 4692, E 28 and elsewhere. It is usually applied to bitumen.

6.2.2   Brittleness temperature

This name is given to the temperature to which a tough or flexible plastic has to be cooled to fracture in a brittle manner when subjected to an arbitrary impact test. D 746 is for materials in sheet form, D 1790 for those in film form. R 974 is a similar type of test in which the sheet under test is bent by the impact through 90° round a 4 mm radius on the jaws in which it is held.

6.2.3   Flammability of solids

A variety of test conditions have been devised to determine whether, on application of a flame, a material:

(*a*)   does not burn, or
(*b*)   burns for a limited time after removal of the flame (self extinguishing) in which case the amount of burning is assessed, or
(*c*)   continues to burn indefinitely, in which case the rate of burning is measured.

In many BS tests a particular form of burner and flame is specified, for example BS 2011, Pt. 2 Pa, but in R 1210 an ordinary bunsen burner is considered adequate. The method of mounting the material varies from one test to another: BS 2782–508A, D 635 and D 229 refer to self-supporting materials which are held with the long axis of the specimen horizontal and the transverse axis at 45° to the horizon, in R 1210 the specimen is entirely horizontal, BS 2782–508B, C and D, R 1326, D 568 and D 1433 deal with flexible sheets or films which are hung vertically or mounted on special frames. BS 3497 is for asbestos board. VDE 0304–3 classifies material in a similar way, but using an incandescent rod at 960°C instead of a flame, this is the same equipment as in the tests described in the next paragraph.

Materials which do not burn or are self-extinguishing by the above tests may be tested for loss of weight when held in contact with an incandescent body, a rod usually of silicon carbide at 950°C, for a specified period, such tests are described in R 181, BS 2782–508E which agrees with it, D 757, DIN 53459 and VDE 0302.

R 871 is a test for the temperature at which inflammable gas is evolved from a solid in a cavity in a heated block.

6.2.4   Pour point, flash point and fire point of liquids

The pour point is an indication of the lower temperature limit of use for a liquid coolant/insulant and is found by cooling the liquid in a jar held vertically until the solidifying material does not move when the jar is turned to the horizontal position: see BS 4452 agreeing with D 97 and BS 148, Appendix E.

The flash point of an inflammable liquid is the temperature of the liquid when the vapour above it will just momentarily ignite— 'flash'—in air when an igniting flame is applied. The fire point is the temperature of the liquid at which the vapour will just stay alight. The flash point is normally used to decide the highest temperature at which it is safe to use an insulating oil, solvent, etc. There are several types of equipment used for these determinations. The simplest is an open cup heated at a controlled rate with a standard flame applied to the liquid surface at intervals: BS 3442, BS 4688 (Pensky–Martens open cup) and D 92 (Cleveland open cup) agreeing with BS 4689 and DIN 51584 specify forms of these.

More precise results for flash point can be obtained using a cup having a lid with an opening which is uncovered only when applying the igniting flame, the best known of these is the Pensky–Martens closed cup, see BS 2839 which agrees with D 93 and BS 148, Appendix C.

When the liquid is viscous at its flash point, or apt to form a skin, an open cup is necessary.

6.2.5   Evaluation of endurance to high temperatures

The temperatures to which materials can be subjected without suffering unacceptable permanent deterioration over the period for which the relevant equipment is expected to survive, when not known by experience, are usually investigated by tests based on principles which are outlined in Section 3.1. The extensive and sometimes prolix literature of this subject is an indication of the industrial importance that has been attached to it, rather than of rapid technical advance, or of the use of difficult concepts or of the attainment of precision. The principles are simple and the methods have generally been known for fifteen or more years. The quality and reproducibility of the data obtained are indifferent (for a combination of reasons inevitably arising from the nature of functional tests themselves and from the variability of materials in industrial use); they scarcely justify the precise statistical methods recommended for analysing them, these methods do, however, help to ensure unbiassed interpretation. The importance of the subject would justify more investigation of the reasons for the diversity of the results (Appleby *et al.*, 1970, Mahon, 1969).

The tests must be regarded as significant in a relative rather than an absolute sense. They are generally more stringent than service conditions so that the predicted failure of a winding insulation system after, say, 20000h at 150°C by no means implies an expectation that a motor incorporating this system operating at 150°C would fail after 2–3 years of continuous service. It does imply that such a system would be more reliable than one whose predicted life was only 5000 hours under the same conditions.

6.2.5.1   Design of text procedures for materials

The overall principles on which thermal endurance tests should be based are outlined in IEEE 98 (formerly 1D), IEC 216, and in somewhat different form in VDE 0304–2. These include consideration of the type of service intended and of the use of conditions such as high humidity, mechanical or electrical stress, mechanical and thermal shock, continuously or intermittently; also the effect of one material on another possibly requiring the testing of simple combinations of materials. Samples should be selected with regard to variability of supply. Exposure temperatures should be three or four in number and should be so chosen that the life at the lowest temperature is 5000h or more, and at the highest temperature 100h or more. The length of cycle between proof tests at each temperature should result in about 10 cycles to failure. A sufficient

number of samples, e.g. 10, should be tested at each temperature
to get reproducible mean results. Great attention to uniformity of
conditions in the ageing ovens is needed; there should be ample
circulation of air and the positions of samples should be interchanged during the sequence of cycles. A new international
specification is being prepared, prescribing methods of evaluation
and statistical analysis in much greater detail than does IEC 216.

The German recommendations in VDE 0304–2 adopt a somewhat different approach from that of IEEE 98 and IEC 216: that
of measuring the change in a particular property of a material
(mechanical strength, electric strength, dielectric loss etc.) with
temperature and time rather than the survival time under a particular régime with periodic proof testing. The lengths of time that
different materials would be expected to last in service can then
be compared from the times required for the relevant property to
deteriorate to some arbitrary limit. Otherwise the general principles
are of similar effect to those of the US and international
recommendations.

Each of these approaches, that of time to failure and that of
measured deterioration, is appropriate in different circumstances.

6.2.5.2 Design of test procedures for systems of materials used in electrical equipment

IEEE 99 (formerly 1E) is devoted to this topic. It suggests that
tests should be done on complete equipment where possible; if not,
models as realistic as possible should be tested and confirmatory
tests made on the complete equipment. The principles suggested
are very similar to those of the preceding section.

6.2.5.3 Statistical analysis and extrapolation of data

The various US standards mentioned below recommend the
normal statistical methods for finding the regression line which is
the best fit of a straight-line relationship between log(life) and
(thermodynamic temperature)$^{-1}$ (eqn. 3.1, p. 29); also that other
statistical quantities should be reported, viz.: the standard (mean
square) deviation of the individual observations from the regression
line, which measures the scatter of results on individual specimens;
the 95% confidence limits of the regression line itself, which
measures the limits between which, statistically speaking, a
regression line determined from a fresh set of observations would
be expected to fall in 95% of cases in which the whole set of
observations were repeated; and a test for whether the observations
suggest a departure from the above linear relationship by an

amount significant compared with the scatter of the observations. Details of these calculations are given in IEEE 101 and some of the other IEEE publications dealing with specific tests which predate no. 101. There has been much elaboration and exposition of these methods; see, for example, Nelson (1971/72).

IEC 216 recommends similar procedures in general terms without giving details of statistical methods.

The use of precise statistical methods, and of the phrase '95% confidence limit' should not induce the belief that the results are rendered more precise thereby. If the results for lives of individual specimens at particular temperatures vary within a ratio of about 2:1, nothing can alter the fact that individual samples will vary to about this degree. Unfortunately, moreover, it seems from experience that the 95% confidence limits deduced from internal statistics of one set of observations usually underestimate the variability of the mean regression line when the complete set of observations is repeated a number of times with seemingly identical material and procedures. Generally comparisons of results from different sources show disagreements of the order of 2:1 in life or 10–15°C in temperature.

VDE 0304–2 and DIN 53446 suggest relatively simple statistical methods. The mean and standard deviations of the group of observations for each temperature are to be found, together with the confidence limits for each group. No particular procedure for extrapolation is recommended beyond plotting log(time) against (thermodynamic temperature)$^{-1}$.

6.2.5.4   Methods of testing thermal endurance of materials

IEC 172 and D 2307 prescribe the method of testing of round, coated magnet wire alone or in combination with an impregnating varnish, using pairs of wires twisted together with a simple jig, dipped if desired in an impregnating varnish, cured and aged according to the foregoing principles (Section 6.2.5.1). This is regarded as a preliminary test for wire enamels, impregnating varnishes or combinations of these. In its original form (AIEE no. 57) this twisted wire test was the first functional test, codified by an official body, incorporating the Arrhenius' Law principle and the statistical methods now common to all the US tests of this kind.

IEC 370 and D 1932 prescribe the testing of flexible varnishes by coating in a standard manner on to glass fabric, and include a bending test with the voltage proof testing. D 1830 is a similar method for complete coated fabrics.

D 2304 deals with rigid insulating materials and the criteria of thermal degradation which may be used.

6.2.5.5   Methods of testing thermal endurance of machine and transformer systems

IEEE 117 is for systems of insulation of random wound motor stators. It describes a type of model ('motorette'), standard methods of producing coils and slot liners and of impregnating, and the types of cycle and proof testing which can be recommended (see Mahon, 1969). Form-wound a.c. and d.c. motors are covered in IEEE 275 and 304.

IEEE 65 is for ventilated dry power and distribution transformer insulation. A model construction is suggested and a method of arranging thermocouples; the ageing temperature is reached by circulating current in the windings instead of placing them in an oven. Humid atmosphere periods and voltage proof tests between windings and sections of windings and to earth are specified.

IEEE 259 and 266 deal with specialty transformers and electronic power source transformers respectively and include vibration tests, thermal shock test at subzero temperature and induced over-potential test among the conditions applied at each ageing cycle.

6.3   Electrical behaviour

We have already noted that electrical characteristics can be grossly influenced by adventitious substances, especially moisture. D.C., resistivity and electric strength (d.c. or a.c.) are highly and often unpredictably susceptible to these effects; dielectric loss, dispersion (change of permittivity with time or frequency) and discharge level can be substantially affected, tracking and endurance to arcing are least affected except in extreme cases of moisture saturation. It will be taken for granted in the following sections that appropriate conditioning must be employed for a long enough period before testing, and during testing, to get reproducible results; unless, of course, the test is intended to assess the condition of the sample.

6.3.1   Volume resistivity and surface resistivity—solids

A three-electrode system is used, the functions of the three being interchanged according to whether volume or surface resistivity is being measured.

BS 2782–202A and B deals with volume resistivity, specifying dimensions of electrodes which are either of mercury in suitable structures, metal foil adhering with vaseline or graphite painted in colloidal suspension. BS 2782–203A deals with surface resistivity using the same electrodes. VDE 0303–3 (which is identical with DIN 53482) covers similar measurements.

IEC 93 and D 257 deal similarly with volume and surface resistivity but without requiring specified dimensions, and giving additional choice of electrodes of metallic paint, sprayed or evaporated metal, conducting rubber or, for compressible materials, metal plates. IEC 345 deals with tests at temperatures up to 800°C.

6.3.2 Volume resistivity—liquids

A number of test cells are in general use, see Baker (1965a), D 1169, IEC 247 and VDE 0303–3, in all of which the main part of the measured volume is between concentric cylinders with a gap of one to two millimetres. Ease of cleaning is very important because resistivity is greatly diminished by contamination; the various designs have different balances of advantage between this feature and others such as simplicity in use and definition of dimensions. Methods of cleaning are described in IEC 247 and D 1169.

6.3.3 Insulation resistance

This is intended as a quick check on quality or dryness using electrodes which are reasily arranged—binding posts or taper pins in holes drilled at specified spacings—but not adapted to calculating resistivity. BS 2782–204A to D, D 257, IEC 167 and VDE 0303–3 (identical with DIN 53482) specify tests of this sort. IEC 345 deals with tests at temperatures up to 800°C.

6.3.4 Permittivity and dielectric loss

Methods of measurement generally are reviewed by Field and by Westphal in the survey edited by v. Hippel (1954), which also contains tables of values for many materials; also by Baker (1965a), Bennett and Calderwood (1971), in IEC 250, D 150 and in IEEE 51; VDE 0303–4 (which is identical with DIN 53483) gives an account broadly similar to IEC 250.

Definitions and/or explanations of terms and symbols in common use are given in these and other places, but in reading earlier and current literature confusion can easily arise from recent changes, from similar names having different strict interpretations, and from literal translation from German to English and vice versa. (The usual French names are the literal equivalents of the

English ones.) A summary is therefore given below, with the approved symbols.

ε or κ: permittivity, meaning *either* permittivity in MKS units (F/m), the present internationally agreed convention; *or* (in older works) permittivity (dimensionless) relative to free space, dielektrizitätskonstant. (The former is obtained from the latter by multiplying by $8 \cdot 854 \times 10^{-12}$.)

ε_r: relative permittivity, dielektrizitätszahl, according to the present international agreement (in older works called simply permittivity or dielectric constant and symbolised by ε or κ).

$\left.\begin{array}{c} \varepsilon' \text{ and } \varepsilon'' \\ \text{or} \\ \kappa' \text{ and } \kappa'' \end{array}\right\}$: real and imaginary components of complex permittivity, in either of the senses above.

ε'': is also called loss factor, loss index (present internationally agreed term), verlustziffer.

δ: loss angle, verlustwinkel, the angle between the current vector for the material under test and that for an ideal lossless dielectric. Should be expressed in microradians, not degrees.

$\tan \delta$: loss tangent, dissipation factor, verlustfaktor.

$\sin \delta$: power factor, leistungsfaktor.

(Note that for a low loss dielectric $\delta \simeq \tan \delta \simeq \sin \delta$.)

At power and audio frequencies measurements are normally made with a bridge network [see Hague (1971) for the general theory of bridges]. High voltage measurements have generally been made with a Schering (resistance–capacitance) bridge; this suffers from errors due to stray capacitance to earth, which may be made negligible in certain circumstances, or, more safely, eliminated by an additional Wagner earth network (Churcher and Dannatt, 1926; Rayner *et al.*, 1930). In this form the bridge is capable of high precision; the earth network may be replaced by a cathode follower circuit to maintain the detector terminals at earth potential. More recently the availability of high permittivity low-loss steels has made more practical the use of bridges having the windings of a transformer as one pair of arms. For high voltage measurements a bridge can be used in which these two arms consist of a variable ratio current-transformer, balance to zero net flux in the core (measured by a third winding connected to a high impedance detector) by varying the number of turns in one winding and by suitable phase-balance (loss angle) circuits (Baker, 1965*a* and 1965*b*). This is often referred to as an 'ampere-turns bridge'. Bridges such as these in which the capacitance under test is balanced against a capacitance of similar size by changes in low impedance arms of the bridge are obviously necessary for high voltage measurements; they are also convenient for power frequency measurements at low voltage because they permit the phase-balance to be made simply and with readily available components. In the kHz–MHz range, any network which can provide a null balance can be used by simply substituting the capacitance under test by a calibrated variable capacitance and a suitable resistance network, adjusting these to give the same balance. For this purpose a bridge employing a fixed 1:1 voltage transformer for two of its arms and a T-network of resistors for the phase balance (Lynch, 1966; and BS 4542) is convenient.

There are many bridge circuits in which a pure capacitance substitution can be made, but few in which the losses can be measured accurately, without errors resulting from stray impedances; it is therefore advisable to use only those which have been fully explored and to keep within the frequency range for which they are designed.

Capacitance substitution is usually effected by having the sample between plates whose separation is controlled by a micrometer, on removing the sample the plates are moved closer together to rebalance and the permittivity calculated from the thickness of the sample and the separation of the plates at rebalance.

In the radio frequency range, 20 kHz to 100 MHz, (overlapping with the bridge methods) the substitution method using a capacitance-inductance resonant circuit is general. On removal of the sample the micrometer capacitor is readjusted to resonance. Losses in the two conditions can be determined by measuring the resonance widths, the difference being attributable to the sample. The most generally satisfactory method, using a small incremental capacitor in parallel with the sample capacitor, is that of Hartshorn and Ward (1936) described in BS 2067. Alternatively the ratio of losses with and without the sample can be found by the *Q*-meter method, i.e. by measuring the ratio between the responses of the resonant circuit in the two conditions with the same injected voltage. This is a convenient method for repetition work. A refinement of the resonance method for low-loss materials of relative permittivity near to two is the immersion-substitution method using a low-loss liquid of nearly the same permittivity, described in D 1531. Refinements of the Hartshorn and Ward techniques, including high precision and automatic recording, enabling low losses to be measured with an accuracy within about $\frac{1}{2}$ microradian are described by Reddish *et al.* (1971).

In the more limited range from 500 kHz to 30 MHz a twin-T network may be used; this is a null-balance substitution method, the capacitance of the sample being replaced by a variable air capacitance, as in other methods, and phase compensation being obtained by change of another variable capacitor.

For frequencies above 100 MHz the logical extension of the capacitor and coil is the re-entrant cavity, in which the 'inductance' is a tube surrounding the capacitor (Works, Dakin and Bogg, 1944) and which is usable up to about 500 MHz. Alternatively the sample may occupy a short length of concentric transmission line. At frequencies up to 10000 MHz the commonest methods involve the insertion of the sample into a waveguide with various ways of measuring responses and deducing the properties of the sample. The basic principles of these methods and the calculations involved have been outlined by Westphal (see v. Hippel, 1954), and reviews of recent techniques are given by Harvey (1963) and by Bennett and Calderwood (1971).

D 2520 gives a standard US method for microwave loss measurements at high temperatures. Lynch and Ayres (1972) describe a method of measuring very low losses in the frequency range 10–35 GHz using an immersion substitution method for solids.

A quite different principle from any of the preceding has been used by Hyde (1970) covering the range 10^{-4} Hz to 10^6 Hz (in two bands). This enabled him to construct an automatically recording 'spectrometer' for rapidly measuring permittivity and loss tangent over several decades of frequency. The principle consists in applying a step-function voltage of very sharp front and 'sampling' the current at intervals thereafter. The shortest sampling interval is 100 ns after application of the voltage, each subsequent interval increases by a factor of two over the previous one. We have noted in Section 4.1 that the response to an alternating voltage can be derived from the response to a steady one provided these responses are linear and superposable (this is implied by the existence of permittivity and loss values independent of voltage). Hyde's apparatus carries out an approximation to the Fourier transform for the complex permittivity

$$\varepsilon'(\omega) - j\varepsilon''(\omega) = \int_0^\infty i(t) \exp(-j\omega t)\, dt$$

$i(t)$ being the current per unit field strength at time t after the voltage step was applied. This method requires much more elaborate and expensive equipment than the more common ones.

6.3.4.1 Measurements on solid samples

The normal type of two-terminal sample-holding capacitor (Hartshorn and Ward, 1936) has already been mentioned. Where a suitable electrical network is used, e.g. Schering bridge, or transformer bridge, it is preferable to have a three-terminal system, the 'earthy' electrode being surroundeed by a guard ring at the same potential. This defines the field of the guarded electrode, reducing the edge correction, or if the technique suggested by Lynch (1966) is used, eliminating it altogether, enabling more accurate measurements of permittivity to be made. At low frequencies accurate measurement of loss usually requires the use of conducting electrodes in contact with the surface of the sample everywhere. This is to ensure that the guard ring intercepts any surface leakage current round the edge of the specimen, and that currents do not flow in resistive paths across the faces of the specimen, contributing irrelevant losses. Such electrodes are usually adherent films, of metal foil attached by a low loss grease, evaporated or sputtered metal, or in the case of ceramics, sprayed or burnt-on metal coatings; an arrangement of mercury containers can also be used. These are described in BS 2782–202A and B, IEC 93 and D 150.

Various techniques in the preparation of samples and electrodes are included in the standards mentioned at the beginning of this section and in BS 2782–205A to D, –206A to D and –207 A to C (sheets and tubes), D 669 (measurement parallel to the surface of sheets), D 2149 (ceramics up to 500°C), D 1082 (mica) and IEEE 83 (paper tapes for power cables).

6.3.4.2 Measurements on liquid samples

At low and medium frequencies liquids can be tested in the same cells as are used for d.c. resistivity measurements, see Section 6.3.2. For accurate measurements on bridge circuits those which have guard electrodes are to be preferred. Electrode systems of the micrometer type such as the Hartshorn and Ward, with a rim on the lower electrode to contain the liquid, are necessary at high frequencies; above 100 MHz the waveguide methods are adapted to holding liquids.

6.3.5 Dispersion

In general use this term denotes the decrease of permittivity with increase of frequency, which occurs when the material has any dielectric relaxation times near to the half-cycle times of the frequencies involved (Section 4.1).

Often the term is used in the limited sense of a measurement made by a dispersion meter (Mole, 1953) which, in effect, measures the change of capacitance of a sample (often paper) measured by charging or discharging for two different periods. This change is expressed as a fraction of the capacitance corresponding to the longer period. The sample is charged at a constant voltage V_0 for a time t_c, short-circuited for a much shorter time t_d, then open-circuited; the voltage which then builds up is observed with a high resistance electronic voltmeter until it passes through a peak V_1 which it does after a time comparable with the original charging time. The dispersion is for this purpose defined as $D = V_1/(V_0 - V_1)$ and is roughly the same as the fractional change of capacitance between measurements at two angular frequencies $1/t_c$ and $1/t_d$.

This measurement is generally used for checking the dryness of insulation (Constandinou, 1965a and 1965b), and since prominent relaxation times associated with moisture in fibrous insulation are of the order of 1 s the pairs of values used commonly are

$$t_c = 300\,\text{ms and } t_d = 3\,\text{ms, or } t_c = 3\,\text{s and } t_d = 0\cdot1\,\text{s}$$

Fig. 34 shows examples of the effect of moisture on dispersion in paper.

6.3.6 Breakdown voltage and electric strength

Electric strength (unlike, say, permittivity) appears to have no definite value which can be approached more closely by greater refinement of techniques; on the contrary it depends on nearly every circumstance of testing as well as on the material tested. The test methods which have been used are consequently nearly as numerous as the investigations, ranging from those designed to avoid discharge in the contiguous medium to those designed to encourage it, from those aiming at a uniform and precisely measurable field strength to those intended to give the maximum degree of non-uniformity, from time scales of nanoseconds to time scales of days. It is impossible here to summarise the methods that have been used.

This section therefore deals only with the arbitrarily standardised methods of test used for material assessment and control, methods which permit discharge in testing solids and a normal degree of contamination in testing liquids; these being considered appropriate in relation to normal service conditions.

Standards broadly relevant to these tests are BS 923: 'Guide on high voltage testing techniques', IEEE 4 'Techniques for dielectric tests', IEEE 51 'Guiding principles for dielectric tests' and IEEE 82 'Impulse tests on insulated conductors'.

6.3.6.1   Electric strength of solids

Though the general principles underlying all the national standards are the same there are many differences in detail particularly in the electrodes used.

BS 2918 and IEC 243 outline the types of equipment needed for providing the high voltage and define the electrodes to be used: for sheet material (brass cylinders, the upper one 25 mm diameter and the lower 75 mm), for tapes (the upper one a rod of 6 mm diameter cut off square, the lower one a plane) and for flexible and rigid tubes. It also prescribes methods of testing parallel to the surface (or laminae) of sheet, board and tube. The procedures for raising the voltage are defined for proof testing and for determining breakdown voltage on a 'rapidly applied', '20 second step-by-step' or 'one minute' basis; also BS 2918 gives instructions for determining a time-voltage curve. Similar methods are indicated in BS 2782–201 (plastic sheets and tubes), BS 2572 (phenolic laminates) BS 3953 (glass fabric laminates) and BS 3497 (asbestos boards).

Waxes and compounds are solidified around electrodes consisting of 12·7 mm diameter spheres, 1·25, 1·00 or 0·70 mm apart.

The corresponding US standard D 149 specifies pairs of identical electrodes, of a variety of sizes, and time sequences for the application of voltage differing for different materials.

VDE 0303–2 (which is identical with DIN 53481) provides a variety of electrode shapes: disc-plane with four different diameters of disc from 25–100 mm, sphere–plane, sphere–sphere, needle and Rogowski electrodes. The procedures for raising the voltage are comparable to those of the British and international standards.

6.3.6.2   Electric strength of liquids

IEC 156 and VDE 0370 specify brass electrodes of 36 mm diameter with their opposing surfaces shaped as caps of spheres of 25 mm radius, 2·5 mm apart at the axis. A protective circuit interrupting the supply within 20 ms of breakdown is specified. The same oil filling may be tested six times with gentle stirring between.

D 1816 uses similar electrodes to IEC 156 but only calls for interruption within three to five cycles after breakdown. Six tests per filling are allowed. An impeller stirrer within the cell is specified.

BS 148, now under revision to accord with the IEC specifications, specifies brass sphere electrodes of 12·7 mm diameter, 4 mm apart. The sample must withstand the proof voltage for one minute. No interruption of breakdown current is specified (though this is frequently used: Baker, 1965*a*) and the cell is refilled for each test. IEC 156 also allows 12·7 mm spheres as an alternative to the electrodes mentioned above, but they are 2·5 mm apart.

D 877 specifies disc-shaped electrodes of 25·4 mm diameter, 2·54 mm apart with interruption after three to five cycles.

All the standards specify methods of cleaning the cells between fillings; it is extremely important to avoid introducing dust or fibres apart from any already in the sample to be tested.

6.3.7 Resistance of boards to burning by power arcs

These tests are intended for materials for arc-chutes, fuse holders etc., where arcing may impinge on the surface.

BS 738 specifies a method in which an arc of 10 A drawn between carbon electrodes on the under side of the board being tested is maintained until the board is penetrated, the time required for this being a measure of the quality of the board. This test also appears in BS 4145 (glass-resin boards) and BS 3497 (asbestos cement boards).

BS 738 also describes a fuse-wire test in which a copper wire, clamped between two boards of the material under test, is fused with a current of 100 A followed up by an arc maintained by a 500 V d.c. supply, the arc being continued for 5 s after the blowing of the fuse which takes 10 s. The test is repeated until the board becomes conducting or 10 tests have been carried out. BS 3497 describes a modification of this method in which the initial fusing current is provided by a 12 V battery and the follow arc by a transformer, avoiding the need for a high power d.c. supply.

VDE 0303–5 (identical with DIN 53484) describes a test in which a 10 A arc is drawn between carbon electrodes resting on the upper side of the board being tested and the condition of the board is reported.

6.3.8 Susceptibility to tracking

Tracking implies permanent deterioration of a piece of insulation, usually the formation of a conducting track resulting from intense, highly local heating of the material (see Section 4.5). It does not mean merely a flashover. Tests for general charring by a power arc are described in Section 6.3.7, the present section deals with tracking in the most usual sense of the formation of relatively fine conducting tracks associated with moisture and/or dirt. It differs from the effects of partial discharge, corona etc. in that high temperature degradation is involved, and it can occur at voltages less than 300 V.

BS 3781 describes the determination of the Comparative Tracking Index (CTI). Electrodes 4 mm apart rest on the horizontal surface of the specimen and drops of an 0·1 % solution of ammonium chloride fall from a hypodermic tube at the rate of two per minute. The electrodes are supplied at variable voltage up to 500 V, 50 Hz. Tests are made at various voltages until failure (the tripping of a relay at 0·5 A) occurs or 100 drops have fallen without failure. A plot of the number of drops required for failure against voltage is made and the CTI is the numerical value of the voltage at which an infinite number of drops would be expected just to produce failure. If failure by tracking does not occur the depth of the erosion pits is measured.

IEC 112 uses the same apparatus but defines CTI as the voltage at which 50 drops just produce failure; which is signalled by the tripping of a relay set at 0·1 A. Parkman (1961) has examined some of the subsidiary processes and the effect of electrode and electrolyte materials in this type of test.

VDE 0303–1 prescribes alternative tests. One is similar to BS 3781 but differs in several details, notably that the 'kriechstromfestigkeit' is determined by the highest voltage at which failure does *not* occur in any of five successive tests. The alternative test is to determine the number of drops required to cause failure at 380V a.c., or if 101 drops do not cause failure to classify the material in one of three classes according to the depth of the erosion pits.

The US tests are numerous and elaborate; there are four ASTM methods, two intended as short-time tests and two which take several hours, and are intended to reproduce the condition for tracking caused by atmospheric moisture. The shorter tests are D 495—high voltage low current dry arc resistance, and D 2303—differential wet tracking resistance. In D 495 a current, initially of 10 mA, from a 12·5 kV supply is sparked between electrodes resting on the sample for $\frac{1}{4}$s every 2s over a period of 60s. During subsequent minute-long regimes the on–off ratio is increased until it is continuous, and then the current is increased, finally reaching 40 mA over the sixth and last minute. The 'arc resistance' is measured by the number of seconds of this sequence the material survives before it becomes conducting, this is the figure usually quoted in tables of material properties. In D 2302 the inclined specimen dips into ammonium chloride solution and currents from a 3 kV source are applied between the liquid and a small electrode $\frac{1}{16}$ in away until this gap becomes conducting.

One of the longer tests D 2303—liquid contaminant inclined plane tracking and erosion—is the nearest to BS 3781 in principle but the specimen is tested on its under side, which is supplied with ammonium chloride solution. Conditions of increasing severity are applied by increasing the flow rate and voltage according to a schedule. Either the voltage at which tracking begins or the time to track one inch, or both, may be used quantitatively. If there is no tracking the time and voltage to produce a specified depth of erosion are noted.

The other elaborate test is D 2132—dust-fog tracking and erosion test. The specimen is sprinkled with a specified dust, wet by water from a fine spray, and a voltage increasing from 500V applied between electrodes $\frac{1}{2}$ in apart. The subsequent control of voltage is complicated, the condition which causes conduction or, failing this, the final degree of erosion is noted.

Billings *et al.* (1968) have compared the dust-fog, inclined plane and IEC tests applied to materials which are highly resistant to tracking, and pointed out that the US tests are increasingly tending to compare something related to erosion; the numerical results may be irrelevant to tracking in those materials which do not readily form carbonaceous conducting tracks. Kurtz (1971) has also published a comparison of the first three above US methods applied to epoxy materials together with a test using the Tracking Endurance Wheel, a device in which samples are sprayed at intervals of a minute or so with water of resistivity a few thousand ohm-cm, 10–20 kV being applied between electrodes six inches apart. The above comment of Billings *et al.* would seem to apply to this test also.

6.3.9 Discharge detection and measurement

The simplest method of detecting discharge in high voltage bushings, to listen for the hissing noise, is not satisfactory, because it is subjective and can be upset by other noises; nevertheless, it has been successful. Its possibilities for spatial location, where other methods of location fail, and still interesting, see for example Haraldsen and Winberg (1968).

An upturn of the loss-tangent/voltage plot of a cable, bushing etc. has long been known as a useful, if imprecise, warning of the onset of discharge. A less ambiguous method was developed by Arman and Starr (1936) in which the high-frequency electrical noise from the discharge is separated and measured. Discharge in the sample is distinguished from noise from the high-voltage source or elsewhere by incorporating the specimen in a resistance-capacitance bridge balanced at a frequency corresponding to a major harmonic component of the discharges. A filter circuit of pass band centred on the same frequency (e.g. 10 kHz) is connected to the detector points of the bridge and feeds an indicating instrument. Then electrical noise reaching the bridge along the supply lead is attenuated, but noise originating in the test object itself, forming one arm of the bridge, appears at the detector terminals. The later developments of this bridge using a c.r.o. detector are described by Baker (1965*a*) who estimate that it can be sensitive to about 10 pC on a test object of capacitance a few hundred picofarads.

Modern methods of excluding extraneous noise are more effective, and a c.r.o. display on which the pulses can be individually counted and measured is normally used. The general features are indicated in Fig. 35; they are discussed by Mason (1965) and in BS 4828, agreeing with IEC 270, and D 1868. VDE 0434 (1966) is more restrictive in prescribing the apparatus to be used.

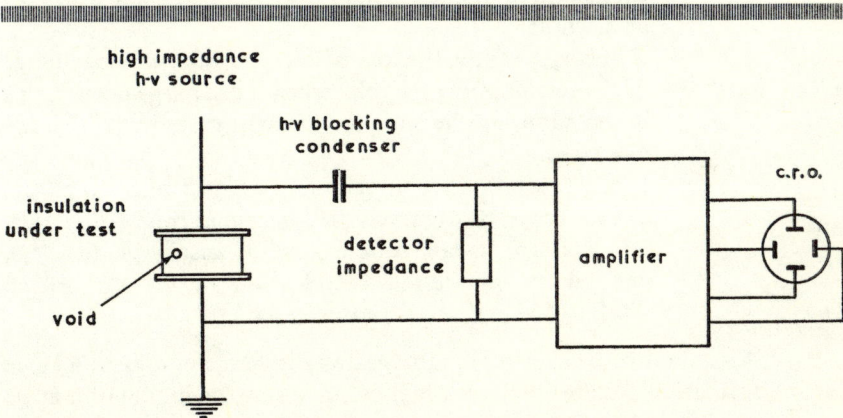

Fig. 35 Basic circuit components for discharge detection

Two types of detector developed by the Electrical Research Association are outlined by Mole (1967) and compared by him in some detail (1962). The narrowband type, in which the voltage supplied to the amplifier and c.r.o. is that which appears across a resonant circuit, is easier to shield from extraneous noise by means of filters in the supply circuit, and is suited for measuring inception and extinction voltages and for counting pulses provided they are not too frequent. Since resolution along the time axis varies inversely as bandwidth it is not suitable for circumstances where numerous discharges are to be investigated or where it may be necessary to discriminate between a directly transmitted pulse and a reflected one; as may happen in investigations on transformers or cables. For such purposes a wideband detector circuit and suitably noise-free high voltage source are needed.

The quantities normally measured with a discharge detector are:

(a) *Inception voltage* (V_i) at which discharge first appears when the voltage is being raised.

(b) *Extinction voltage* (V_e) at which discharge ceases when the voltage is being lowered.

(c) *Discharge magnitude* or *discharge level* (*Q*), found by measuring the normal peak fluctuation of voltage δV at the terminals of the test object and multiplying this by the capacitance of the latter. This may be roughly related to the actual quantity of electricity carried by the discharge from a knowledge of the relative dimensions or capacitances of the discharge site and of the test object as a whole.

(d) *Discharge energy*, the loss of energy from the capacitance of the test object, $\frac{1}{2}QV_i$ pk., when a discharge occurs, this is the same as the amount of energy dissipated by the discharge.

(e) If the resolution of the circuit does not allow separate discharges to be observed the *quadrature rate of discharge*, in effect the mean square rate of discharge in C^2/s, derived by square-law rectification of the unresolved discharge signal, is sometimes used.

A method of measuring the integrated discharge energy in a large insulation system (such as a high voltage stator winding of a large machine where the discharge energy may be around $10\,\mu J/pF$ with a great number of discharge sites) is described by Simons (1964). The detector signal from a capacitance bridge, one arm of which is the insulation under test, is fed to one pair of plates of a c.r.o., a fraction of the voltage supplied to the bridge being applied to the other pair. When the bridge is balanced in the absence of discharge the c.r.o. displays a straight line, but if the voltage is raised to the inception level the line broadens out into a rhomboidal loop, the area of which is a measure of total discharge energy per cycle. Dakin and Malinaric (1960) use a similar bridge to find the total discharge magnitude and suggest that the slope of the c.r.o. trace, as a measure if the increase of capacitance of the specimen as the voids become short-circuited by discharge, can be used to estimate the total volume of voids (but one must make assumptions about the shapes and orientations of voids to do this).

Discharge measurements on transformers and other high-voltage gear are often made with equipment designed for determining the noise signal which will be produced in radio communication receivers (radio influence voltage), these are not readily interpretable in terms of discharge magnitude, but are simple and often give useful comparisons.

Discharge during impulse voltage applied to insulation can be measured by a method described by Mitra, Sakr and Salvage (1965).

6.3.9.1   Interpretation of discharge measurements

Where the test object can be regarded as a lumped capacitance, interpretation may be simple, but where it is a distributed impedance, such as a transformer winding, the signal arriving at the terminals will be more or less attenuated and modified according to its point of origin.

Methods of locating discharges in transformers have been widely discussed and are generally based on recognising three components of the signal arriving at the terminals:

(a)   The travelling wave component, propagated along the conductor with a velocity comparable to that in free space; this is only detected in simple windings, not, for instance, in interleaved disc windings.

(b)   Capacitance transfer between discs in a disc winding.

(c)   Low frequency oscillation of the windings as *LCR* networks.

A single discharge site giving a component of type (a) may be located from the difference of arrival times at the two ends of the winding.

A signal with a strong component of type (b) may be located if there is access to the winding for injection of impulses at various points to compare with those due to the discharge.

Signals of type (c) may be identified in some cases with particular regions of origin.

The measurement, interpretation and location of discharges in transformers is receiving considerable attention; see, for instance, several of the Committee 12 papers of CIGRE 1968, Kawaguchi and Yanabu 1969, Austin and James 1970.

Measurement of discharge in large power capacitors presents problems, arising mainly from internal inductances, which are discussed by Mole, Parrott and Kendall (1969).

A survey of discharge test methods for cables, and of methods of scanning cables for defects has been made by Kreuger (1966).

6.3.10 Endurance of materials to partial discharge

This subject has already been discussed in Section 4.3.3, here we will merely indicate the standards that have been established. All of them refer to the possible use of frequencies higher than power frequency but advise that checks be made at power frequency. IEC 343 offers two warnings about this, that dielectric heating may cause earlier breakdown and thus too short an apparent discharge life, and that the more rapid rate of discharge may result in more rapid release of degradation products which may either accelerate breakdown or form conducting layers suppressing discharges and delaying breakdown.

IEC 343 prescribes electrodes consisting of a flat plate as the lower one and a vertical rod 6 mm diameter, cut off square with rounded edge, as the upper one. With soft material the rod should be held up to $100 \mu m$ away from the specimen. Tests should be conducted in flowing dry air of not more than 20% relative humidity. New materials should be tested at three voltages such that the shortest life is about 100 h, and the longest 5000 h equivalent 50 Hz life. Routine tests should aim at a voltage to give 1000 h equivalent 50 Hz life.

VDE 0303–7 and DIN 53485 specify electrodes similar to the IEC ones but there is no rounding of the edges. Instead of measuring the number of cycles to breakdown materials may be compared by measuring properties after various periods of discharge.

D 2275 specifies an upper electrode in the form of an upright cylinder $\frac{1}{2}$ in diameter with rounded edges, or a $\frac{1}{2}$ in diameter ball bearing, resting on the sample under a pressure of 90 gf. Ambient air of 25% or 50% relative humidity is recommended.

6.4 Water absorption test

The very simple procedure of immersing a weighed conditioned sample of (usually) specified dimensions in water for a stated period and observing the increase of weight on removal has a number of variants. Some standards specify boiling water for some materials and cold water for others, some leave a choice, various periods of immersion are specified, the longest in common use is 24h. The following is a representative selection of standards prescribing tests of this kind: BS 2782–502A to G, BS 2782–503A to C, R 62 (equivalent to DIN 53475), R 117, DIN 53471, DIN 53472, D 570, BS 2572 all referring to plastics; BS 4145 (glass-bonded mica), BS 1598 (ceramics).

In the US and some British standards the results are expressed as percent. weight/weight; but many British standards require the result as the total absorbed in milligrammes. Care is necessary in comparing these results because different dimensions are specified for different types of materials, in some cases the thickness of sheet material is specified, in others it must be stated.

Where the material under test may contain substances which can be leached out by the water, the specimen is weighed after immersion and again after drying to constant weight.

Tests for absorption of water from atmospheres of controlled humidity also exist.

6.5 Water vapour transmission or permeability

BS 2782–513A to D, E 96 and R 1195 describe similar types of test in which a film of the material being tested is sealed around the mouth of a hollow cup which contains a desiccant (or water in E 96). The cups are placed in a cabinet of controlled humidity and temperature and the gain of weight (loss if water is inside) observed over a period. Results are usually given in g/m^2h for a stated thickness. The passage of water is affected by the physical state of the film, e.g. crystallinity, and divergent results are often obtained for this or other reasons. The simple diffusion laws are frequently inapplicable.

6.6 Determination of small moisture contents

Unless a large sample is available or the moisture content is high the simple process of weighing, heating and weighing again is not sensitive enough. All methods suffer from uncertainties due to hysteresis and the difficulty of defining the 'dry' condition, particularly for cellulose materials.

The methods given here are intended for solids, especially paper. Methods for oil are given in Section 10.1.

E

6.6.1 Vincent–Simons method

Described by Vincent and Simons (1940), this is a direct method in which the sample is placed in a container at −70°C which is then evacuated and closed. It is heated to 100°C and the equilibrium vapour pressure measured. The water vapour thus released is absorbed in phosphorus pentoxide and the process repeated as often as necessary. The total water content is computed from the successive vapour pressures and the volume of the container, using an extrapolation formula.

6.6.2 Karl Fischer method

The reagents and procedures are described by Mitchell and Smith (1954) and in BS 2511, D 1533 and R 760. The Karl Fischer reagent consisting of iodine, pyridine and sulphur dioxide dissolved in methyl alcohol reacts quantitatively to remove moisture from methyl alcohol or a methyl-alcohol–chloroform mixture in which the moisture-bearing substance has been washed or dissolved. The standardised reagent is added in measured amount until the point at which all moisture has been used up is indicated, either roughly by a colour change, or accurately by the onset of current flow between two platinum electrodes at 20–100 mV, which cease to be polarised when water is absent. The application to paper is described by Flanagan and Jenny (1972) and by Waddington (1959), who describes a titrimeter for detecting minute quantities of moisture.

6.6.3 Phosphorus pentoxide electrolytic hygrometer

The sample is heated in a stream of dry gas which then passes over a bifilar spiral winding of platinum wire embedded in a layer of phosphorus pentoxide on a quartz rod (Still and Cluley, 1972). Voltage applied between the two legs of the winding causes current to flow electrolytically when water is absorbed, the current continues until all the water has been electrolysed, thus the total charge passed measures the water that has been absorbed.

6.6.4 Electrical properties

Dispersion or loss tangent (Constandinou, 1965*b*) are simple to measure and probably more reliable than conductivity, but, as discussed in Section 5.1, they are not very reliable as quantitative measures, particularly at moisture contents below 1 per cent; in some circumstances, however, no other method is possible. Constandinou (1965*a*) considers that the ERA dispersion meter (Section 6.3.5) can give reliable results.

Chapter 7

Traditional fibrous and solid insulants

The materials considered in this chapter were, until about 30 years ago, practically the only ones in common use (apart from electrical porcelain which is included in Chapter 11), and are still among the most widely used. They largely formed classes A and B of the old thermal classification (Section 3.2). The literature about them is very extensive, and the object of this chapter is to remind the reader of their main characteristics and to indicate sources of information about them. Many, indeed most of the materials, are the subjects of specifications issued by the standardising bodies named at the beginning of Chapter 6; these can normally be found from the indexes of these standards under the name of the material or class of material concerned.

7.1 Cellulose paper, pressboard and vulcanised fibre

These, the cheapest of fibrous materials, and the simplest to apply, are used for their flexibility and simplicity of handling and shaping when the working temperature is not too high; they also provide moderate strength combined with lightness and low cost in resin bonded boards, tubes, bushings etc.

Cellulose is a long-chain molecule, a polysaccharide, and one of a number of vegetable substances formed from repeated glucose units

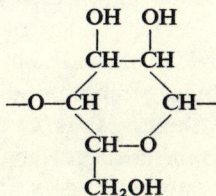

$$
\begin{array}{c}
\text{OH} \quad \text{OH} \\
| \qquad | \\
\text{CH—CH} \\
\diagup \qquad \diagdown \\
\text{—O—CH} \qquad \text{CH—} \\
| \qquad \diagup \\
\text{CH—O} \\
| \\
\text{CH}_2\text{OH}
\end{array}
$$

The molecules usually consist of a few thousands of these units, and form a mixture of crystalline regions (micelles) in which there is some degree of order, and amorphous regions in which the chains form a random tangle. The gross structure consists of fibres 15–30 µm diameter and 1–20 mm long.

Cellulose fibres used for electrical papers are largely derived from coniferous woods pulped by the kraft process, the basis of which is boiling chips of wood in a superheated alkaline liquor. (This is also called the sulphate process because loss of caustic alkali from the liquor is periodically made up by adding sodium sulphate, evaporating, calcining and boiling with lime.) Fibres of cotton, flax and manilla are also used, generally obtained from rags and old cordage which are also broken down by boiling in alkali. The suspension of fibres is thoroughly washed and then 'beaten' by passing between fixed and rotating blades or bars and finally matted on travelling gauze and dried on travelling felt.

The essential requirement for electrical papers is purity, especially freedom from resins, gums, and a variety of substances known collectively as lignin which are broken down by the alkali process, and from alkalis and salts which are removed by the washing processes. A description of the types of cellulose fibres, paper-making methods and types of paper in the context of electrical usage is given by Warren (1931, Chapter 24). At the time that was written it was not considered practicable to obtain wood fibres of sufficient purity for the best electrical papers, but with modern processing the source of fibre is not important from this point of view. There are nevertheless very characteristic differences between fibres from different sources, for example cotton fibres contain the highest priportion of pure α-cellulose (as distinct from related substances) while manilla fibres are among the strongest.

The outstanding factors in choice of papers are:

> mechanical strength (tensile, bursting and tearing)
> density
> impermeability and porosity
> uniformity

Mechanical strength depends on processing variables, and on types of fibre, and is usually greater the longer the fibre. High density and impermeability, provided they are obtained in the right ways, tend to increase the impulse strength and also the permittivity of oil-impregnated paper; these results can be obtained by more intensive beating and shorter fibre lengths, generally at some expense of mechanical strength. A smooth finish and higher density, but not, according to Kelk and Wilson (1965) much greater impermeability, can be produced by calendering (passing over hot rolls at the end of the manufacturing process). Fairly dense, smooth papers are commonly called 'grease-proof'.

Papers in much greater variety than can be considered here are produced to try to obtain the optimum combination of properties required in manufacture and in service for various types of equipment. Several types of creped papers are now produced which can be used for taping conductors and joints of irregular shape, or current transformer hairpins, with greatly reduced risk of tearing, but at the cost of a somewhat lower impulse strength for a given thickness of insulation.

Esterified papers, particularly cyano-ethylated paper which has a better endurance to high temperature than natural cellulose, and acetylated paper which has lower moisture absorption than cellulose, are in regular use. Various proprietary treatments are believed to improve the temperature endurance of papers significantly.

Experiments with papers combining synthetic fibres with cellulose ones, composites of paper with plastic fibres or with layers of exfoliated mica have been made; the latter has been suggested for oil-impregnated cable dielectric having a lower dielectric loss than paper itself.

7.1.1 Behaviour in use of oil-impregnated paper

The properties of unimpregnated and impregnated papers are discussed at length by Clark (1962) and more briefly by Hall and Kelk (1962), Kelk and Wilson (1965) and King and Wentworth (1954). Well-dried lapped oil-impregnated papers as used in high voltage cables may have a loss tangent less than 2×10^{-3} at 20°C and 3×10^{-3} at 100°C with a slight minimum between. Permittivity varies with type of cellulose and density, the relative permittivity being usually in the region of 3–3·5 (that of the fibres themselves is about six). The dielectric loss appears to be partly associated with the proportion of amorphous cellulose (in the region of a few per cent, this part is assumed to be dipolar), partly with the small quantity of residual sodium ions present, the latter contribution appearing at high temperatures (Barry, 1970). This is rather confirmed by the observation that at the higher temperatures the loss tangent of paper capacitors impregnated with trichlordiphenyl diminishes a little with increasing stress in the region of 1–20 V/μm (Garton effect, see Section 4.2.3.3).

At low temperatures Allan and Kuffel (1968) found that the loss tangent of nearly dry paper rises to a peak of 0·01–0·02 at 160 K, 50 Hz (240 K at 90 kHz) and then falls steadily to about 3×10^{-4} at 4·2 K. Damp paper behaves in generally similar fashion but with higher peaks in the 160–140 K region and somewhat lower value at 4.2 K. Permittivity was recorded as close to 1·60 from 100 K downwards, in both cases.

The impulse strength, often the limiting factor in determining the thickness of high voltage cable insulation, can be improved by up to 25% by higher density of the paper, and by up to 10% by higher viscosity of the impregnant; but the higher permittivity accompanying increased paper density increases the capacitance current to earth, while viscous impregnants are more prone to allow voids to develop during service than mobile ones. High density paper also takes longer to dry. In commercial cables impulse strengths of 100–150 kV/mm can be achieved.

Other important considerations in a.c. oil–paper cable design are low loss (to reduce heating, increase efficiency and, in the case of very high voltage cables, avoid thermal breakdown) and low level of discharge at working voltage to ensure long life. The relative importance of these factors changes with voltage; at very high voltage ratings the impulse voltage requirements become very important, but also reduction of loss tangent and avoidance of discharge become very crucial at the high working stress necessary. At distribution voltages, on the other hand, the precautions necessary at high working stresses to prevent discharge would be unwarrantably expensive; much lower stresses are used. In modern practice layers of semi-conducting (carbon-impregnated) paper are taped on the stranded conductor to form a smooth electrostatic screen for the overlying paper insulation, and prevent electric stress in the spaces between the surface strands. The field strength in the insulation is therefore that between two concentric cylinders, the greatest intensity being at the surface of the semi-conducting screen:

$$E_{max} = \frac{V}{a \ln(b/a)} \qquad . \quad . \quad . \quad . \quad . \quad 7.1$$

where V = working voltage, a = radius of inner screen, b = outer radius of insulation. However E_{max} is not necessarily kept the same for different ratios b/a even at the same voltage rating, and in some ranges

$$E_{mean} = V/(b-a)$$

remains more nearly constant. For 'solid' impregnated distribution cables used up to 33 kV a.c. rating (19 kV conductor to sheath), BS 6480 prescribes the same (or nearly the same) mean stress at any particular voltage rating, ranging from 2·3 kV r.m.s./mm for 11 kV systems (6·35 kV conductor to sheath) to 2·7 kV r.m.s./mm for 33 kV systems, VDE 0255 prescribes slightly lower values. The corresponding stresses at the conductor are higher by 50% and 80% respectively for the smallest conductor sizes and by 10% to 20% for the largest. These are about the highest a.c. stresses that can be used with this type of cable which is rarely free from discharge after a period of service. At the other extreme, in e.h.v. cables filled with low viscosity oil under pressure mean stresses of 6–7 kV r.m.s./mm for 230 kV and 300 kV systems, and 9–11 kV r.m.s./mm for 400 kV systems (with stresses of 10–16 kV r.m.s./mm at the conductor) have been operated (Barnes, 1966). At such voltages it is usual to base the insulation design on stress at the conductor, and it is the impulse test voltage which is the limiting factor rather than the working voltage. It has been suggested that a cable for a 750 kV system could be designed with a working stress of 20 kV r.m.s./mm at the conductor (Ball *et al.*, 1972).

High voltage d.c. cables are usually required only for submarine or urban power links, and the number of these is small. Dielectric loss is irrelevant and discharge does not become a problem until stresses are much higher than can be used for a.c. cables of similar type. The England–France 100 kV d.c. and the Italy–Sardinia 200 kV d.c. cables, of 'solid' impregnated paper construction, have mean working stresses of 14 kV/mm and 17 kV/mm respectively (Oudin *et al.*, 1967). The effects of temperature and stress on conductivity near the conductor tend to relieve the high stress which eqn. 7.1 (applicable to d.c. only if conductivity is uniform) predicts. This, however, implies a space charge in the body of the insulation near the conductor, so that impulses of opposite polarity to the working voltage will produce a very high stress in this region. Various authors have suggested working stresses of 35–40 kV/mm for oil-filled d.c. cables.

In impregnated paper capacitors the dielectric consists usually of a number of thin layers of paper (thicknesses down to about 5 μm are available, though expensive, 10–25 μm is common) impregnated with mineral oil or chlorinated diphenyls. The latter have relative permittivities similar to cellulose fibres, i.e. about six, with a significant dielectric loss (see Section 10.2), stresses of 15–18 V/μm are commonly used. The types of paper employed and the number and thicknesses of layers depend on the voltage and a variety of other factors (Barry, 1970). The use of three or four layers of paper ensures that a defect in one layer does not result in breakdown of the dielectric at that point. Many paper capacitors have metallised film electrodes; if a breakdown through the dielectric occurs the concentration of current at this point evaporates the metal and effectively isolates the fault.

Impregnated paper capacitors are largely used for power factor correction of inductive loads and fluorescent lighting circuits, and in electronic equipment for blocking and bypass functions in situations where very low loss is not essential.

Oil-cooled power transformers are insulated almost wholly with paper; lapped tape on the conductors, pressboard collars, cylinders and sheets for inter-coil, inter-layer, inter-winding and winding-to-earth insulation (see, e.g., MacDonald, 1954). The general requirements are similar to those for cables, but field configurations with both impulse and power frequency voltages are complex and factors affecting discharge, such as thickness of insulation, are very varied. Analysis of the electric fields in high voltage transformers is a normal part of their design, mainly to locate regions of abnormally high stress, the normal working stresses are considerably lower than in cables of similar rated voltage. The difficulties of identifying and locating discharge sites in transformers (see Section 6.3.9.1) has prevented the accumulation of design experience based on discharge studies; hitherto the overvoltage test of 2·8–3 times rated voltage for 1 mm has been effective in ensuring good service life.

Oil-impregnated 'condenser' bushings (using buried foils to control the stress distribution) present a more manageable but by no means simple field problem. The designer can achieve a uniform axial field, or a uniform radial field, or some other compromise. In high voltage bushings Marlow (1970) suggests that radial stresses in the region of 3·5 kV/mm can be safely used. This implies considerably higher local stresses, but at this general level of stress some degree of discharge is allowable (see Section 4.3.6). Axial stress is determined by flashover in the external medium, air or oil, of course.

Operating temperatures for oil-impregnated paper are usually limited to 100–110°C. Allowance for hot spots in transformers and for emergency overloads in cables may mean that the general temperature is not normally greater than 70–80°C. It may be that better design of cooling systems and allowance for higher emergency temperatures will allow normal transformer operating temperatures to be raised.

7.1.2   Pressboard and vulcanised fibre

Pressboard is prepared from cotton rag fibres or from pulp processed like other electrical papers, but instead of drying out as a single layer, a number of wet layers are placed together, pressed in a hydraulic press and dried by heat. In general behaviour it is, of course, similar to paper, though of somewhat higher density; its electrical properties are inferior to those of layers of lapped paper of similar total thickness. It is used for interwinding barriers, spacers etc. in transformers; impregnated with resins it is used for a variety of purposes such as slot liners, spools, barriers, end cheeks etc.

Vulcanised fibre (which goes under a variety of names such as leatheroid, fish paper, horn fibre) is made by treating thick paper with zinc chloride and hydrochloric acid and winding on a mandrel so that the gelatinous partially dissolved layers of cellulose fuse together; then leaching out the zinc chloride, drying and calendering. It is thus a partially reconstituted cellulose, and forms hard tough sheets or tubes of varying grades. It can be worked to some extent like soft metal, its electrical properties are inferior to good paper, but it is used for a large number of components which are punched, drilled or moulded and require toughness combined with low-grade insulating properties. It is widely though perhaps decreasingly used for slot liners in low voltage machines, barriers in switches and many applications where a cheap and tough material is required to operate under dry conditions with little or no electric stress. It has some degree of flame resistance and can be used for arc channels on contactors and switchgear. Normally it contains several per cent of moisture.

Laminated, resin-bonded paper boards, tubes etc. are dealt with in Section 9.1.

7.2   Natural fabrics

Cotton, linen and jute fabrics are declining in use for electrical purposes as glass, terephthalate and other non-absorbent fabrics and various kinds of polymers fill the purposes which they used to serve. Coarse cotton and jute fabrics are used for making tough laminated boards, cotton tapes are still in fairly general use but cords and sleevings are disappearing. Silk has been wholly displaced. The use of these fabrics before the advent of newer material is described by Warren (1931). The properties of cotton fabric laminates are described and tabulated by Clark (1962) and by Oburger (1957).

E*

7.3 Derivatives of natural cellulose

It is obvious from the structure of cellulose (Section 7.1) that its susceptivility to moisture and its thermal stability are likely to be improved by replacing the three hydroxyl groups of each glucose unit with more stable, less lyophil groups.

The esters and ethers of cellulose are not so 'traditional' as the other materials of this chapter, but mostly they pre-date the wholly synthetic thermoplastics considered in Chapter 8. They are relatively cheap, light and strong but they are much more susceptible to moisture than most thermoplastics (though much less so than paper); their thermal stability is not much better than that of cellulose, and they have substantial dielectric loss even when dry.

By treating cotton fibre or wood pulp with a mixture of acetic acid and acetic anhydride cellulose triacetate is formed. This polymer is used as such (see below) but for many purposes it is partially hydrolysed by adding water to give cellulose acetate, which normally means an ester intermediate between diacetate and triacetate. Cellulose acetate-butyrate and cellulose propionate are produced analogously. For most purposes these polymers have to be plasticised, and a wide range of materials, mostly organic esters, is used. The most important outlet for cellulose acetate is the production of fibres for fabrics, but these are little used in electrical applications. The plasticised esters can be extruded, injection moulded, cut into sheets, or cast into films down to 10μm thick.

The density is around $1 \cdot 1$–$1 \cdot 3$ g/ml. Mechanical and electrical properties are very dependent on the amount and type of plasticiser used, the following figures indicate the range to be expected: tensile strength at yield 15–40 N/mm^2, elastic modulus 300–3000 N/mm^2, Rockwell R hardness 70–120, permittivity 3·5–5 and loss tangent in the range 0·01–0·07 whether measured at power, audio or radio frequencies. Water absorption (24 h) may be up to 4 % for cellulose acetate, about half this for the acetate-butyrate; they are affected by many organic solvents, acids and alkalis. Resistance to tracking is good, but not to discharge. Elongation at rupture can be up to 60–70 % for the acetate, greater for the acetate-butyrate.

Mouldings can be used in electrical equipment where the insulating function is not demanding, temperatures do not exceed 70–80°C and moisture is no problem. Films are used for insulating tapes, interlayer insulation, liners etc. and when protected with suitable varnishes can be used at up to 105°C (Class A). Cellulose triacetate is used as a film for wire and layer insulation with somewhat better temperature stability than the other esters.

Ethyl cellulose is an ether made by treating cellulose with caustic soda and then with ethyl chloride to convert about 80% of the hydroxyl groups to ethoxy groups. It has properties in the same range as the esters but it requires less plasticiser, is more flexible and has greater toughness, especially at temperatures down to −40°C, and has somewhat lower dielectric losses. It has some tendency to cold flow.

7.4 Asbestos

The group of fibrous minerals known collectively as asbestos have needle-like crystal structures, and on the atomic scale consist of chains of silicate and other metal oxide radicals. The most commonly used one, *chrysotile*, is considered to have a unit structure represented by $Mg_6Si_4O_{10}(OH)_8$. Other asbestos minerals belonging to the *amphibole* group, include *amosite* with the approximate composition $R_6(Si,Al)_7O_{22}(OH)_2$ (R may be Ca, Na, Mn, Fe, Mg), *crocidolite* which has a somewhat different arrangement of the same metal atoms and *tremolite* with the empirical formula $Ca_2Mg_5H_2(SiO_3)_8$. Chrysotile fibres, which are the more suitable for spinning, are up to a centimetre or two in length and very thin, down to 20 nm in diameter.

The principal sources of supply are in South Africa, Rhodesia and Canada. The mineral is crushed, teased, beaten with steel paddles and after removal of foreign material, screened and mixed with reinforcing fibres if required. The fibres are then carded, i.e. combed into parallel orientation, and twisted into roving, or wound as lap on to rolls. Yarn, cord etc. is produced from the roving; cloth tapes and sleeving are woven from the yarn. For electrical use the choice of asbestos free from iron, and the removal of any iron oxide or metallic particles is important. Woven materials may be reinforced with glass or natural fibres.

Asbestos paper and mat are made by methods similar to those for cellulose paper.

Chrysotile fibres begin to lose water and deteriorate noticeably at about 450°C but suffer some loss of strength at 300°C; prolonged heating in air at temperatures even below 200°C will ultimately produce loss of strength (Clark, 1962). They are, of course, much better than cellulose fibres at temperatures above 100°C, but worse than glass fibres at temperatures above 200°C.

Amosite is more stable to heat than chrysotile and tremolite more stable still. None are affected by oil but they are attacked by strong acids. All forms of asbestos absorb atmospheric moisture very noticeably, probably on account of the very large surface area of the fibres, rather than any special affinity of the substances for water.

Asbestos materials are not suitable for high voltage insulation; nor for high frequency applications because of their high dielectric loss even when dry. Their main electrical use is in low-voltage high temperature situations and for confinement of arcs. As a reinforcement or filling for phenol formaldehyde laminates or mouldings asbestos reduces the tendency to moisture tracking.

Chrysotile fibres are more readily spun when cotton or other organic fibres are incorporated, and woven materials may contain up to about 20 % of organic material, though pure asbestos products can be obtained when heat stability takes precedence over mechanical strength. Rovings with a reinforcing core of cotton, glass or other fibres can be obtained. Asbestos paper may contain glass or other fibres, it is very weak unless treated with a resin bond.

Wires for high temperature situations are covered by serving with asbestos lap or roving, or by using sleeving.

Asbestos boards are made in a variety of ways. The cheapest are bonded with starch and calendered; these are not very strong but will stand fairly high temperatures. Resin bonded papers and boards from about 0·1 mm thick upwards use, for example, phenol formaldehyde, polyvinyl acetal, epoxy or silicone resins according to the temperature of service.

A detailed account of the properties of asbestos products is given by Clark (1962).

Laminates can be made from felts or woven cloths with appropriate high temperature resins. They are used for low-voltage Class H transformers, armature slot wedges, furnace parts, domestic heating equipment and similar applications.

7.5 Mica

The most obvious characteristics of the micaceous minerals are strength in two directions and weakness in the third; so that they split readily (more or less so according to the type of mica) into plates or sheets, which may be extremely thin. The two types of most interest electrically are *muscovite* and *phlogopite*; they are found in all the continents but the majority of good electrical mica comes from India. Muscovite is the best and most widely used for electrical purposes. Micas are formed from sheets of complex silicate structure consisting of layers of oxygen atoms grouped around silicon, aluminium or other metal atoms, and weakly held together by layers of widely spaced potassium atoms. The main features of the atomic structure of muscovite are shown in Fig. 36 (Plate); phlogopite is similar but with magnesium replacing some of the aluminium and silicon. Typical basic characteristics are: density 2·6–3·3 g/ml, hardness (Moh) 2·5–3·2.

The best electrical micas are clear and transparent with varying degrees of coloration (muscovite generally ruby or green, phlogopite amber); poorer grades have inclusions or spots which may represent electrical weaknesses and are often dark coloured. There are standard systems of grading for mica according to appearance.

Muscovite is stable at temperatures up to 500–600°C, phlogopite up to 800°C. Above these temperatures the OH groups in the structure decompose with the production of steam and disintegration takes place.

Good muscovite has a loss tangent typically in the range 10^{-3}–10^{-4} over a wide band of audio and radio frequencies at room temperature. Loss tangent and conductivity increase rapidly at high temperatures and this may limit the maximum usuable temperature. Relative permittivity generally lies between 5·5 and 7. Electric strength in thicknesses of 0·1 mm or less is $1–2 \times 10^5$ V/mm. 'Intrinsic' strengths as high as 10^6 V/mm have been found.

One of the most valuable characteristics of mica is its resistance to erosion by prolonged discharge, which is much greater than that of any other flexible material.

Synthetic mica can be made by heating potassium silicon fluoride (K_2SiF_6) with silica, magnesia, alumina to form a melt from which the mica can be slowly crystallised. These processes were developed, particularly during the Second World War, in the hope of replacing natural mica in countries which had difficulty in obtaining it, but it has never competed in price with natural mica. Its properties are very similar to those of natural mica, but the OH groups in the structure (see Fig. 36) are replaced by fluorine atoms; it is therefore stable at higher temperatures.

7.5.1 Uses of natural sheet and flake mica

At one time mica was widely used for low-loss high frequency capacitors, coated with silver it is still so used but is being superseded by high permittivity ceramics and low loss polymer films. Its low loss and mechanical properties make it suitable for microwave windows between evacuated and atmospheric regions. Flakes are also used for interlayer insulation in Class H transformers, for winding non-inductive resistors and potentiometers and for mounting the components of vacuum tubes.

7.5.2 Built-up mica flake materials

Mica splittings of large area are becoming scarcer, and for most purposes it is uneconomic to use flakes larger than 3–5 cm diameter. These are built into sheets or plates by laying them in overlapping configuration with an adhesive between the flakes, often on a paper, fabric or glass cloth backing. The flakes may be laid by hand or by machine, coated with liquid resin, or scattered from a tower together with a powdered resin which melts when the sheet enters the curing oven. A great variety of resins may be used according to whether the final product is to be rigid, flexible, suitable for hot forming, slit into tape etc., and the temperature at which it is to be used. These sheets can be made up into tubes, liners, spools, or wrapped or taped on to conductors etc.

7.5.3 Mica paper

The increasing price of splittings of reasonable size and the manufacturing convenience of using uniform and homogeneous materials has led to the development, during the last two decades, of paper-like materials made by exfoliating the smaller mica flake material into very tiny plate-like particles about $1\,\mu m$ thick, and forming them into sheets by settling from suspension in a machine similar to a normal paper-making machine. (No cellulose or other fibres are used.) Exfoliation is effected by heating to about 800°C to commence disintegration, followed by either: (*a*) quenching in water, mechanical disintegration and screening, or (*b*) quenching in sodium carbonate solution and treating with dilute sulphuric acid so that the carbon dioxide forces apart the laminae and completes the disintegration. Another process employs high pressure jets of water to break up the mica mechanically (Chen and Staley, 1967; Ketterer, 1964).

The cohesive force between the platelets is sufficient to give the mica paper the strength to permit gentle handling, but its tear strength is too low for most purposes until it has been treated with a binding resin. This, of course, limits the usable temperature to the maximum which the resin will stand; the resin also affects the dielectric loss. There is no particular limit to the thickness of mica paper than can be produced, but it is generally available from $50\,\mu m$ to $\frac{1}{4}$ mm thick.

The resistance to discharge of mica paper appears to be comparable with that of flake mica (McKeown and Olyphant, 1965) but direct comparisons are scarce and contrary views have been stated (Noren and Ball, 1969). It is thought to be impaired if resin penetrates between the mica particles.

Combinations of exfoliated mica with other materials to give high temperature and discharge endurance have been described: e.g. in a mixed paper with aromatic polyamide fibres (Smyser, 1969), or sandwiched with polyimide film (McDonald, 1969) or impregnated with silicone elastomer (Rembold, 1964).

7.5.4 Uses of mica flake and mica paper products

Most of the products formerly made from built-up mica sheets can be reproduced from mica paper, resin impregnated and bonded under pressure and heat (Ketterer, 1964). Both types will be considered here together.

Insulators between commutator segments are made up to the required thickness with the minimum of resin. V-rings for commutators, slot liners and other shaped parts are made from hot forming sheet, which is bonded with a soft resin so that the mica layers can slide over one another and conform to the shape of the mould.

Rigid plates are used for winding high-temperature heaters such as those for domestic toasters and flat-irons (Doyle, 1969). In the former type vitreous bonds or silicone resins, which are reasonably stable at the operating temperature, must be used. In the latter type the element is under compression in service and the bond need not survive the operating temperature.

Glass-bonded mica products (Dubois, 1966) are generally made from powdered mica with a borosilicate glass as bond, they have some of the properties of ceramics but can be accurately moulded at temperatures around 600°C.

An important use of mica (whether as built-up flake or mica paper) is for the insulation of high-voltage machines, notably the main insulation of generator stator windings. In many cases Class B or F insulation is necessary because high-voltage machines are often designed for high temperatures, but even where this is not so (where the conductors are water cooled, for example) mica is used because of its endurance to discharge. It is not practicable entirely to avoid discharge in or around solid insulation applied to an assembly of rectangular strands, with Roebel transpositions, fitted into a slot in steel laminations, from which it protrudes at the ends. The use of hydrogen as the gaseous cooling medium in modern large generators has not removed the problem of discharge erosion, though it may have eased it (Breitenstein, Johnston and Maughan, 1969). Where the temperature is high but the voltage is not, on the other hand, or where d.c. is concerned, other high temperature materials are being used instead of mica; generally mica is used on machines operating at 3kV a.c. and above.

Fig. 37 shows a cross-section of one side of a coil of a typical generator stator. The main insulation is made by winding on the conductor stack flexible tape made of built-up mica or mica paper, backed with thin glass fabric, to a thickness of about 4mm for an 11kV machine. (Alternatively the bar may be wrapped with sheets of mica insulation and taped only at the ends where bends occur; Denham *et al.*, 1972.) Thus the nominal stress at working voltage is 1·6kV rms/mm, the maximum stress at corners and slot ends being considerably greater. Laboratory experiments (Davies, 1970) suggest that higher stresses could be used.

mica tape bonded with shellac, bitumen, polyester or epoxy resin, backed with silk, polyester or glass

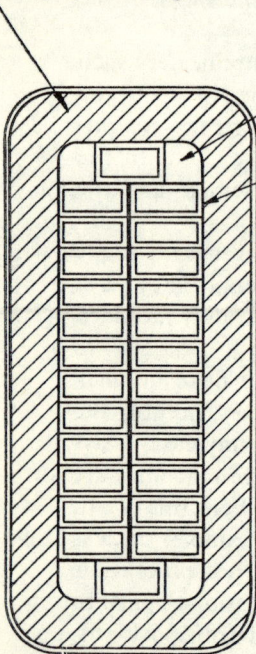

epoxy mica packing material

glass braid or asbestos roving

asbestos coated with a conducting medium

Fig. 37
Cross-section of slot portion of typical alternator coil, showing section of conductors (top and bottom) in course of Roebel transposition
(Farmer, 1970)

The tape or sheet may be bonded with oil–bitumen varnish and the finished coil impregnated with bitumen (Jones, 1952) but on long machines this is no longer the practice because difficulties have occurred due to longitudinal movements of the tape ('tape separation' or 'girth cracks') arising from the difference in expansion between copper and insulation. These difficulties became apparent in large machines in the 1950s, a recent summary is given by Kuzawinski and Wolff (1970).

Modern insulation for high-voltage machines, particularly the larger ones, is usually produced by one of three methods (Farmer, (1970)

(a) Mica tape backed with woven glass is lapped on to the conductor stack, the assembly is vacuum dried and impregnated with an unsaturated polyester-styrene formulation (see Section 9.4), then held in a jig while it is cured by heat, (Laffoon *et al.*, 1951; Anderson, 1965; Fergestadt and Schanche, 1968).

(b) Similar to (a) but impregnation is by epoxy resin (Section 9.5) of low reactivity held at a temperature at which it has low viscosity but does not cure rapidly (Blinne *et al.*, 1961; Rembold, 1964; Matsunobo *et al.*, 1972; Rogers, 1967). This process has also been applied to complete stators (Mertens *et al.*, 1967).

(c) Liquid resin is applied liberally during taping, or the tape is impregnated beforehand with epoxy resin in a volatile solvent and dried to a flexible B-stage (Section 9.5) in which it can be handled or machine-wound on to the conductor, as illustrated in Fig. 47, but contains a negligible amount of solvent. After taping the bar is hot pressed to consolidate and partly cure it, the excess of resin being extruded together with trapped air; a post-cure follows removal from the press (Flynn *et al.*, 1958; Erdman and Lauroesch, 1969; Parriss, 1971). Kohn *et al.* (1971) outline a process in which resin-filled tapes and subsequent vacuum pressure impregnation are combined.

Each of these processes presents its own difficulties in eliminating voids due to imperfect impregnation, trapped air, or trapped solvents.

The formulation of the resin is a compromise between a number of factors such as strength to resist tape separation, low loss tangent at high temperature to avoid thermal breakdown and sufficient flexibility to accommodate the distortion of the end windings on short-circuit test and to avoid fatigue due to vibration, as well as the manufacturing needs of one or other of the processes just mentioned. Fig. 38 (Plate) shows the completed stator of a machine insulated with epoxy-bonded mica.

Similar insulation is also used on large electromagnet coils for high-energy particle accelerators.

7.6 Bitumens, waxes and compounds

7.6.1 Bitumens and pitch

Bitumen and asphalt are commonly, though perhaps incorrectly, treated as synonymous terms. They are impure hydrocarbon materials from mineral sources, presumably of vegetable origin; some such as gilsonite, grahamite, wurtzilite and impsonite are mined in solid form, others are derived from petroleum distillation residues. The term pitch is normally used for a solid black material resulting from wood or coal distillation.

As with most naturally occurring substances there is a good deal of variation between bitumens. Cold flow occurs in most of them although they may be brittle to impact; flow points up to 175–200°C are obtainable. Flow point may be raised by 'air blowing' at 200–300°C, which produces a rubbery consistency.

Bitumens provide a satisfactory cheap medium for moisture sealing but are not self supporting so can only be used as coatings, impregnants and fillings. They are used, often with addition of drying oils, for vacuum impregnation of motor and generator coils, and for sealing of capacitors. With suitable fillers they are used to make cheap mouldings such as battery boxes and terminal boards. Modified by, or incorporated in natural oil drying varnishes they are used in black baking varnishes, as adhesives for built-up micas and for impregnating fabric tapes. They impart flexibility to varnishes and retard drying. They are partly soluble in oil; they have high dielectric losses.

Bituminous materials are used (but decreasingly so) for filling cable joint boxes and terminations for impregnated paper cables.

An account of the sources of bituminous material is given by Warren (1931) and by Smith (1969) who also describes bituminous varnishes and paints. Their general properties and electrical behaviour are described by Clark (1962).

7.6.2 Waxes

There are a number of naturally occurring waxes notably ozokerite (from which ceresin wax is derived by refining) and montan wax; but the majority of wax in electrical use is derived from petroleum residues. Paraffin wax is a hydrocarbon with straight or branched chain molecules of 20–40 carbon atoms; it contains small amounts of naphthenic and aromatic impurities. It is coarsely crystalline, melts usually in the range 50–55°C, has a relative permittivity of 2.2 and low dielectric loss; indeed pure paraffin wax probably has a dielectric loss as low as any known material. The waxes known as microcrystalline (or 'amorphous'), obtained by solvent extraction from pretroleum residues, have a higher proportion of non-paraffinic compounds and a melting temperature from 60–90°C.

Waxes are soluble in oils and many solvents and are mechanically very weak, but are very good moisture sealants and are useful where temperatures are not high, provided they are well supported. They are subject to oxidation at elevated temperatures like the hydrocarbon oils, after an induction period they develop acidity and dielectric loss.

They contract considerably on solidifying, with a tendency to leave voids. Conversely on heating to near the melting point they tend to 'bleed'.

7.6.3 Compounds used in medium-voltage impregnated cables

The impregnant used for 'solid' paper-lapped cables must be sufficiently fluid at about 120°C to impregnate effectively, yet sufficiently viscous at a reasonable operating temperature, say 80–90°C, to prevent movements and 'draining'. A combination of a high viscosity mineral oil with rosin (see Section 7.7.2) has been in use for many decades and much cable with this impregnant is still in service. Rosin is cheap and imparts the necessary viscosity, but the amount that can be used is limited by the dielectric loss it produces with consequent reduction of the current rating of the cable. Its phenanthrene molecular structure gave it the advantage of absorbing any hydrogen produced by the effects of internal discharge on the oil, thus preventing voids from growing, but the use of a proportion of naphthenic oil can fulfil the same function. Compounds based on microcrystalline wax (Section 7.6.2) with a melting range near or above the desired operating temperature (Brazier, 1954), became alternatives to rosin compounds for several years, these have now been displaced by naphthenic oils with a hydrocarbon polymer such as polyisobutylene of moderate molecular weight, (see Section 10.4) incorporated as a thickener.

7.7   Natural drying oils and resins

The three basic functions of impregnants in solid insulation (mechanical consolidation, sealing of cellulose materials from moisture ingress and avoiding gaps and cavities in high voltage insulation) could not be fulfilled merely by materials which solidified by cooling and remelted on heating; for this would have meant either that impregnation would be carried out at a temperature that would degrade cellulose materials or that the impregnant must soften at a temperature below 100°C so that service temperatures would have been limited to this low value. The methods and materials of the paint and varnish industry were therefore borrowed to provide impregnants which could be permanently solidified by heating, and for the first third of this century the oleoresinous varnishes were the most important impregnants available (Warren, 1931); they are still used in substantial quantities. They are based on the vegetable 'drying' oils which polymerise readily to form relatively waterproof films, and on resins derived from tree barks which do not polymerise but which form solutions in many organic solvents from which waterproof films can be produced by evaporation; the resins are incorporated with the drying* oils to produce coatings of improved properties. The introduction of phenolic and alkyd resins was followed by many combinations of the natural with the synthetic materials. This subject now forms a whole complex semi-empirical technology (Martin, 1972; Mills, 1952).

7.7.1   Natural drying oils

Linseed oil, tung oil (also called china wood oil or wood oil), soya bean oil, castor oil and others are obtained by squeezing the various seeds and kernels or by solvent extraction. Linseed and tung oils solidify on exposure to air or on heating, soya bean oil polymerises only partially, castor oil as extracted does not polymerise (in this condition it is used to plasticise varnishes and prevent embrittlement with ageing) but it can be 'dehydrated' by heating at about 280°C to form a readily polymerisable oil.

These oils are generally triglycerides of fatty acids and the

---

* The reader unused to varnish terminology should note that while 'drying' may mean simply the evaporation of a solvent leaving the film-forming resin, as in the drying of shellac spirit varnish, in many contexts it means the conversion (polymerisation) of a natural oil such as linseed oil to a solid by heat, oxidation etc.; it is in the latter sense that the phrase 'drying oil' is used

most important polymerisation process is probably addition polymerisation at unsaturated bonds (Chapter 2); but more complex processes involving oxidation also occur. The oils are refined and treated in various ways before incorporation into varnishes; 'boiled' or 'blown' oils are heated while air is bubbled through them, 'stand' oils are heated to a high temperature with little or no access of air. 'Dryers'—small quantities of catalysts to promote rapid hardening—are usually added, they consist of oil-soluble organic salts (naphthenates, linoleates etc.) of metals such as lead, manganese, cobalt, zinc, some of which modify or intensify the action of others.

Oleoresinous varnishes are widely used as cheap magnet wire covering, for coating fabrics, impregnating windings and sealing generally; they are still very useful in combination with phenolic, alkyd and other synthetic resins to give desirable combinations of properties such as fluidity during impregnation, sufficiently rapid drying, long tank life when used for dipping of components in an open vat, and retention of flexibility at service temperatures. Their electric strength is good, generally they are not particularly susceptible to tracking, but they cannot be used for any purpose where low dielectric loss is necessary.

7.7.2 Natural resins

The natural resins are derived directly or indirectly from the barks of various trees and are typically hard solids, soluble in many organic solvents but insoluble in water and resistant to decay in damp conditions (Barry, 1969, gives a detailed account of them).

Rosin, or colophony, the commonest and cheapest, is obtained by distilling off the essential oils from the exudate of various species of conifer. It consists largely of abietic acid and related acids, the molecules of which contain a single acid group attached to a phenanthrene nucleus (three linked 6-member hydrocarbon rings). Rosin melts at about 80°C; its acidity may be reduced by forming the calcium salt (adding lime) or by esterification. It is not used in good insulating varnishes but has been largely used with mineral oils for the impregnant in low voltage paper insulated cables (Section 7.6.3).

Copal is the name for a number of hard, colourless to brown resins produced by various species of tropical trees in Congo, Zanzibar, the Philippines formerly, Borneo and elsewhere. The harder grades are dug from the ground in forests or former forests (and hence are often called fossil resins), these are usually insoluble and have to be 'run' (heat treated) before use in varnish making.

Damar is rather similar to copal, softer, and obtained from trees native in the Malay Peninsula and Archipelago.

These resins may be used as spirit varnishes merely by dissolving in an alcohol or other volatile solvent which evaporates after application, but their principal use is as a means of improving and controlling the properties of drying-oil varnishes.

Shellac is a secretion of the lac insect, which lives attached to the branches of various trees native in parts of India, sucking the juices from the bark and covering itself with the resin. The encrustations are washed and processed, usually by crude traditional methods, to produce a solid of orange to brown colour (depending on purity), melting at 75–85°C. Its chemistry is indefinite. It is unique among the natural resins in having some thermohardening properties; by heating, its melting range may be raised to about 120°C. Shellac is the most valuable of the natural resins for insulating purposes, on account of its hardness, adhesion and resistance to moisture. It has been widely used as an adhesive bond for built-up mica sheets for wrapping on to conductors of high voltage machies and for making commutator separators and V-ring insulation.

Chapter 8

Linear synthetic polymers and elastomers

Polymers in which each molecule consists of a chain (often, but not necessarily, of carbon atoms) with few, if any, major branches, have a number of features in common. Since the chains are not attached to each other by primary bonds the polymer melts when the temperature is high enough to overcome the secondary forces between molecules but not high enough to decompose the chains themselves. Since they have no reactivity they do not form larger molecules or cross-linked chains on heating so that heating does not itself alter the melting temperature. For these reasons they are called thermoplastic materials.

Some of the polymers have every unit of the chain symmetrical, like polyethylene (Fig. 9), polytetrafluorethylene, polyparaxylylene and polyoxymethylene; these crystallise readily unless they have short branches at irregular intervals. Other polymers described here have an asymmetry at every second carbon atom; thus polypropylene (Section 8.1.3) has one hydrogen and one excrescent carbon atom (with its attendant three hydrogens) on every other carbon in the spine, polyvinyl chloride has one hydrogen and one chlorine similarly disposed, polystyrene one hydrogen and one phenyl group, and so on. These polymers have an additional structural property called *tacticity*; two different regular structures known as *isotactic* and *syndiotactic*, are possible; this is a property of a sequence of several chain units. Isotactic polymers are characterised by the fact that, if the main chain carbon atoms are arranged in planar zig-zag form (like the chain of Fig. 9*a*), the atoms or groups responsible for the asymmetry all lie on the same side of this plane; in syndiotactic polymers they alternate from side to side. These configurations cannot be interchanged merely by rotation about carbon–carbon bonds (a fact which is difficult to appreciate without a three-dimensional model). These regular forms are only obtained when each monomer is attached to the growing chain in an oriented co-ordination with the preceding ones; this is done only by certain special types of catalyst, the process being called stereoregular or co-ordination polymerisation. The majority of monomers not made by these means are *atactic*, i.e. irregular. The importance of tacticity is that atactic polymers cannot be crystallised (unless the atom or group responsible for the asymmetry is very small, e.g. fluorine or hydroxyl). When an isotactic or syndiotactic polymer crystallises, however, it does not necessarily do so in the planar zig-zag form used for convenient definition of the types.

Generally speaking, linear polymers have improved tensile strength, tear strength, low temperature toughness, softening temperature and impact strength the greater the chain length, or, as it is usually put, the greater the molecular weight; but fabrication (moulding, extrusion etc.) is made more difficult because of the high melt viscosity.

The concept of molecular weight of a polymer is not simple; it necessarily means an average, and different measurements give different kinds of average, while different properties respond to different kinds of average. The two simplest are:

(i) the number average, which would be obtained by taking the sum of the weights of all the molecules and dividing by the number of molecules;

(ii) the weight average, which would be obtained by taking each monomer unit of each chain in turn, noting the weight of the molecule of which it is a part, summing these and dividing by the total number of monomer units.

The viscosity of a molten polymer depends on the weight average at low rates of shear, but more nearly on the number average at at high rates. The glass–rubber transition temperature rises with number average molecular weight. Tensile strength is considered to depend on number average rather than weight average, impact strength the converse. Thus the relationship between length of molecular chain and desirable properties is complex and imperfectly understood.

An increase of crystallinity (i.e. of the proportion of chain units lying in crystalline regions) is predominant in increasing density, stiffness and yield point, which are largely independent of molecular weight. It also increases hardness, tear strength, chemical and solvent resistance and softening temperature. A high degree of crystallinity decreases toughness and flex life and tends to produce brittleness and low elongation.

Factors which decrease crystallinity and density are irregularity such as randomly distributed branches and, in polymers with asymmetric chain units, atacticity.

Polymers with an intermediate degree of crystallinity often exhibit a desirable combination of toughness and hardness at temperatures not too much above the glass–rubber transition, and provided the time scale is not too long. The equivalence of time and temperature (Chapter 2) is only roughly applicable to these materials. In some recently exploited thermoplastics, notably high density polyethylene, polypropylene and polycarbonate, crystallinity is a decisive factor in determining mechanical properties; its practical importance is likely to increase.

Many long-chain polymers when above their glass–rubber transition temperatures can be cold drawn: when the tensile stress exceeds the yield stress a very large extension takes place with a small increase of stress, this extension takes the form of a sudden 'necking down' in diameter at a particular region, and the drawing of a thin section from a thick one as though through an invisible die. The drawn-out portion contains molecules predominantly aligned along the direction of drawing, with an increase in crystallinity unless the original material was strongly crystalline, usually it has a higher tensile strength than the undrawn material. Changes of this kind suggest the reason for the non-linearity in the mechanical behaviour of crystallisable polymers and for the variability of their time–temperature relationships.

Many commercial plastics are copolymers of two or more monomers. These may be produced by polymerising a suitable mixture of the monomers, so that the polymer chains consist of the different units in some form of alternation (not usually regular). There are also techniques for producing copolymers consisting of long sequences of each monomer, these are called 'block copolymers' and may have properties different from random copolymers of the same monomers. It is also possible to attach chains of one polymer to various points along an existing chain of another, thus producing branches which are different from the spine, these are called graft copolymers. Copolymers are, of course, amorphous.

Values of the main mechanical properties of the more important linear polymers are given in Table 2; these are for comparative purposes, differences between the values quoted should not be regarded as necessarily significant unless they are large. Thermoplastics are being increasingly used with glass reinforcement, a

Table 2   Typical properties of thermoplastic moulding and extrusion materials

	Density g/ml	Hardness (Rockwell)	Tensile strength N/mm²	Compressive strength N/mm²	Tensile modulus elasticity N/mm²	Elonga- tion %	Impact strength (Izod) J/m	Heat deflection temperature at 18·5, kgf/cm² °C	Burning
Polyethylene—low density	0·91-0·94	R3-15	5-20	—	100-300	100-700	—	33-40	slow
Polyethylene—high density	0·94-0·97	R30-50	15-40	—	500-1200	20-1000	50-2000	45-55	slow
Polypropylene	0·90-0·91	R90-110	30-40	70	1000-1600	200-700	30-300	50-60	slow
Polystyrene	1·04-1·09	M70	40-80	70-100	2000-4000	1-4	10-30	65-105	slow
Polystyrene—high impact	1·04-1·09	M20-70	40-70	30-60	1000-3500	2-80	30-600	65-100	slow
Polytetrafluoroethylene	2·14-2·28	R10	15-35	15	400	100-400	100-300	60	none
Fluorinated ethylene- propylene	2·14-2·17	R25	20	—	300-500	200-300	—	70 at 4·6 kgf/cm²	none
Poly(vinyl chloride)—rigid	2·14-2·20	R10	40-80	40-80	2000-4000	1-5	20-1000	55-80	slow or self- exting.
Poly(methyl methacrylate)	1·17-1·20	M40-100	30-100	70-100	2500-3000	2-40	15-25	70-100	slow
Polyethylene terephthalate (film)	1·37-1·40	—	150-300	—	3000-5000	40-140	—	—	burns
Polycarbonate	1·20	M70-120	50-80	80	2500	60-130	60-90	130-140	self- exting.
Acetal	1·42	M80-100	70-100	130	2000-3500	25-75	80-130	100-125	slow
Polyvinyl formal	1·2-1·3	M80-90	30-100	—	2000	—	30-60	—	burns
Nylon 66	1·10-1·15	R110	60-90	—	1500-3000	60-300	50-100	70-100	self- exting.
Polyimide	1·45-1·50	M110	100-150	130-170	3000	70 (film)	10-60	240-360	none
Polysulphone	1·24	M70	70	100	2500	50-100	70	175	self- exting.

table of the properties of some of these is given in Insulation/ Circuits Encyclopedia.

No attempt has been made to indicate trade names corresponding to the chemical names of the polymers, except in one or two cases where the chemical names are rarely used. There is not always a one-to-one correspondence between trade and chemical names, but fairly comprehensive lists of trade names associated with such polymers are given by Roff and Scott (1971). Their book

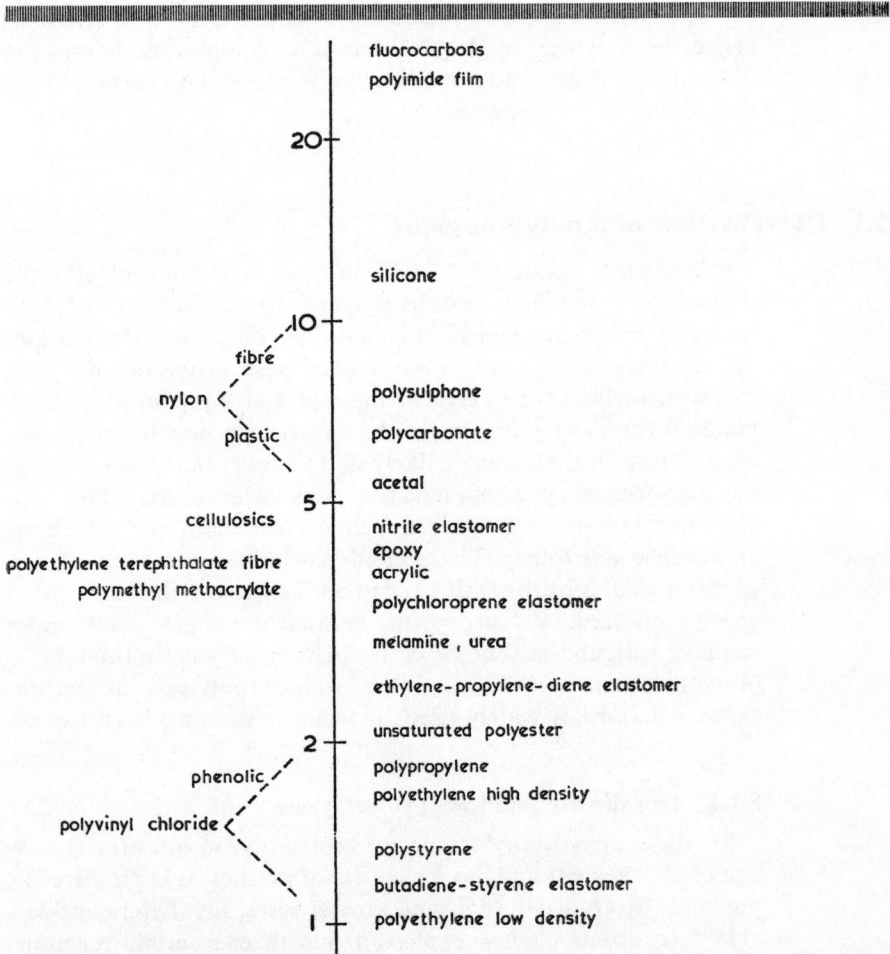

Fig. 39 Approximate relationship of basic prices of various common thermoplastic and thermoset materials
(Based on figures given by Billmeyer, 1970, and by Menges and Dalhoff, 1972)

gives compact summaries of the preparation, properties, processing etc. of each polymer. Other recent textbooks dealing with plastics generally are those by Billmeyer (1970) and Brydson (1969).

There are few occasions when the technical attributes of a polymer alone determine whether it is appropriate for the job in hand. The cost of the material and the cost of processing it in relation to the job are important factors. Although these cannot be assessed out of context, a general perspective of relative costs of materials is useful and Fig. 39 is intended to supply this (cross-linked polymers are also included). Like the table of mechanical properties it is for rough guidance only, comparable figures are difficult to arrive at and it may be that in particular cases the order may become interchanged.

8.1   Polyethylene and polypropylene

Polyethylene produced by addition polymerisation of ethylene (Fig. 4) was the first flexible polymer to approach the 'ideal' electrical properties which, till then, had only been available in hydrocarbon waxes and polystyrene. These properties did not, however, include the degree of thermal and mechanical stability required for most purposes in the electrical power industry, nor was it cheap; it then seemed likely that the only outlet would be in the high-frequency communication and radar fields, where low dielectric loss is of overriding importance, and for which its appearance was timely. The Second World War provided a large, growing and price-insensitive market in these fields and rapid growth ensured. Polypropylene became available much more recently with the advent of co-ordination polymerisation. Poly-(4-methyl pentene-1) is a rather similar polymer, at present expensive and not widely used; it will not be considered further here.

8.1.1   Low density (branched) polyethylene

The type of polyethylene produced from 1936 onwards is now one of the cheapest and most plentiful of plastics. It is produced at pressures up to a few thousand atmospheres, involving considerable precautions against explosion due to exothermic reactions developing. This process gives a polymer which is basically of the single chain type but which has short branches of a few carbon atoms at intervals of a few tens of units along the main chain, about 50 per molecule, and an occasional (less than one per molecule) branch of average molecular length. Typical number-average molecular weights range from 20000 to 50000. The range of molecular sizes is very wide, as indicated by ratios of weight-average to number-average molecular weight, typically between 20 and 50.

Low density polyethylene is generally 50–60% crystalline, the chains in the crystalline region being 'straight' (i.e. the carbon atoms in a planar zig-zag) as in Fig. 3. The crystallites are platelets (lamellae) about 10nm thick, the carbon chains being roughly perpendicular to the planes of the lamellae. The molecules themselves, however, are much longer than 10nm and it appears that considerable lengths of one molecule are folded back and forth like a hank of string, within one crystallite, as well as passing from crystalline to amorphous regions or to other crystallites. The crystal melting point is usually 110–115°C.

Softening temperature (Vicat softening 80–100°C) and susceptibility to oxidation (with consequent increase of dielectric loss and general deterioration) limit the maximum useful temperature to 70–90°C according to circumstances. Polyethylene cables are often designed to run at 70°C maximum continuous, 95°C in emergency overload. Polyethylene is highly resistant to acids and alkalis, it swells but does not dissolve in warm paraffinic oils, slightly so in vegetable and silicone oils; above 100°C it becomes soluble in many organic solvents. It is susceptible to sunlight and cannot be used out of doors unless loaded with about $2\frac{1}{2}\%$ of carbon black, which impairs its properties but still allows it to be used for low voltage cable insulation. For normal conditions it usually contains antioxidants. Its water absorption is extremely low; it does not track with moisture films.

It is not among the strongest of plastics but it is now very cheap and readily extruded over wire or into films, rods or sheets, or injection, compression or blow moulded. Processing temperatures range from 130–230°C. Sheets from 25–250 μm thick are available; they can be joined by welding and coated on to metal. The film can be cold drawn.

It suffers from a slight tendency to stress cracking (i.e. long-term cracking or crazing under small tensile stress, associated with attack by traces of solvents or by oxidation). To overcome this, copolymers with 5% of 1-butene, $CH_2{=}CH{-}CH_2{-}CH_3$, or cross-linked polyethylene are used for cables.

Cross-linked polyethylene can be produced by incorporating peroxides and applying a high temperature, after extruding, to bring about vulcanisation. It becomes a soft rubber instead of melting at temperatures above 110°C and retains its form to about 130°C. This is one way of overcoming stress cracking on covered wires; it also prevents the retreat of the material from a soldering point. The maximum continuous conductor temperature allowable is about 90°C, emergency temperatures may be 130°C, short circuit 250°C.

Cross-linking can also be effected by high energy radiation, provided the material is thin enough to be penetrated by the radiation. Such material is used for tapes and tubes, and can be made heat-shrinkable by stretching or by restraint during irradiation.

The electrical behaviour of this polymer has been more fully explored than that of any other. Many studies of electric break-down, both discharge free ('intrinsic') and under discharge, have been conducted on polyethylene because of its convenience and purity (see Section 4.3.1).

The highest short-time, discharge-free electric strength that has been observed is about 0·9 MV/mm at room temperature; in some circumstances breakdown may be the result of electromechanical collapse. The failure of some polyethylenes by slow treeing when immersed in water has been noted in Section 5.1.

The relative permittivity of low density polyethylene is between 2·2 and 2·3, and is accurately related to density d by the expression

$$(\varepsilon' - 1)/(\varepsilon' + 2) = 0\cdot 325d$$

based on the classical Clausius–Mosotti formula of elementary dielectric theory (Barrie, Buckingham and Reddish, 1966).

Dielectric loss has been of great interest for practical reasons of reducing attenuation in high frequency cables, as well as the scientific ones of accounting for the small losses that have not been eliminated.

The room temperature loss angle of commercial material is often given as 100–200 μrad over all frequencies, but this is not the best that can be done. Fig. 40 shows the values obtained by Barrie, Buckingham and Reddish (1966) (see also Figs. 18 and 19).

The small residual losses may be due to a number of causes such as the small (hypothetical) moments of branch points, or of individual —CH_2 groups, or unsuspected impurities, or oxidation. The direct effect of the small quantities of antioxidant needed for stability in processing was eliminated in this work.

Loss at very low temperatures has been investigated by Fallou and Bobo (1970), Vincett (1969) and others. Vincett reports a peak loss angle of 3–4 μrad at 4·2 K, 4 kHz and a peak of similar height

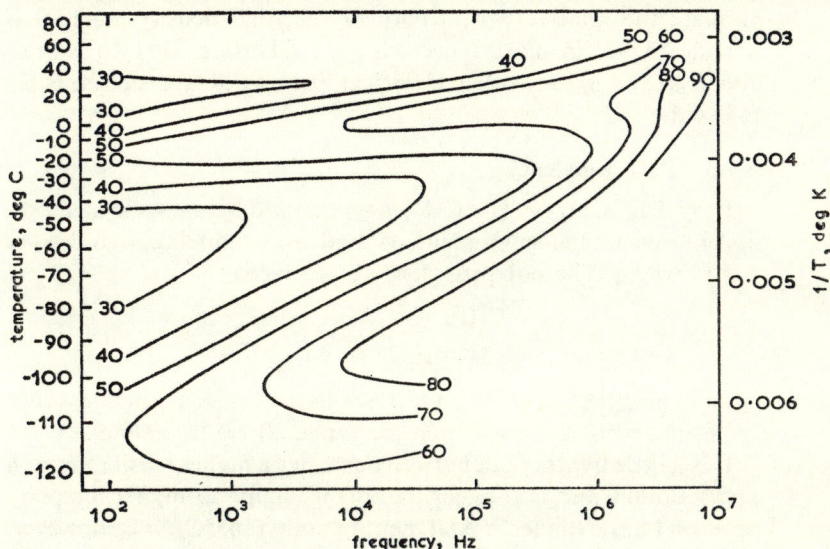

Fig. 40 Loss angle contours as function of frequency and temperature for polyethylene
(Barrie, Buckingham and Reddish, 1966)
This map is on the same principle as the relief model of Fig. 20. Figures marked on the curves are in microradians

at 2 K, 2 kHz. Fallou and Bobo (1970) find about 40μrad at about 4 K, 1 kHz, a minimum of 4μrad at 80 K, 1 kHz, rising to a peak of 100μrad at 160 K and returning to 30μrad at room temperature.

Allan and Kuffel (1968) report loss angles independent of frequency from 4·24 to 80 K of about 30μrad at 1 kHz and 12μrad at 5 kHz.

8.1.2 High density polyethylene

Physically this approaches the 'ideal' polymer more closely that the low density type. It has a very few branches (less than 1 per 200 of the spinal carbon atoms) and is typically 90 % crystalline. It has been produced since 1957 from ethylene by catalytic processes devised by Zeigler and by Natta, and generally known by their names (see Billmeyer, 1970). It is mechanically stable to a higher temperature (crystal melting point 130–135°C), is stiffer and stronger than the low denisty polymer, and has better low temperature brittleness. Zeigler polythene may have a number–average molecular weight up to 200 000.

Apart from these the low and high density forms are generally similar; the relative permittivity of the high density material is a trifle higher in accordance with the Clausius–Mosotti expression already given, other electrical properties are not sensibly affected.

8.1.3 Polypropylene

Propylene, CH_3—CH=CH_2, is synthetised by the same methods, and indeed in the same plant, as used for producing high-density polyethylene. The polypropylene so produced

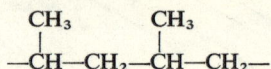

$$\begin{array}{ccc} CH_3 & & CH_3 \\ | & & | \\ -CH-CH_2- & CH-CH_2- \end{array}$$

is isotactic (Section 8.1) and crystallisable. The number-average molecular weight is usually in the range 80000 to 500000.

It is a little lighter than polyethylene, has a higher tensile strength and modulus and is tougher, retains its shape to higher temperatures and has a higher crystal melting point (165°C). It is, however, more susceptible to oxidation by heat or ultra-violet light, and requires antioxidants to ensure stability during processing and service. Pure polypropylene undergoes a glass transition at about 0°C for time scales of the order of 1 s, nearer room temperature for impact tests, and its use is mainly restricted to filaments and certain special applications. Block copolymers of propylene with ethylene are used for injection moulding, and random copolymers for films. These can be processed in the same ways as polythene at temperatures from 210–275°C, and produced as films as thin as 8 μm and in extremely fine filaments. A number of polypropylene 'papers' consisting of polymer fibres, with or without cellulose, have been produced.

Long-term mechanical properties have been investigated by Ogorkiewicz (1970a).

Copolymers of 60% ethylene and 40% propylene monomers (e.p.m.) cross-linked with organic peroxides have properties very like natural rubber. Three-component polymers (terpolymers) of ethylene, propylene and a diene such as 1,4 hexadiene, CH_2=CH—CH_2—CH=CH—CH_3, form rubbers vulcanisable with sulphur (e.p.d.m. elastomers). These rubbers can be used as cable insulation at conductor temperatures of 90°C continuously and at much higher temperatures in emergency. They do not have brittle temperatures higher than −75°C.

The electrical properties of polypropylene are generally similar to those of polyethylene, the dielectric loss differs in detail, having

Fig. 1
Model of sodium
chloride crystal
structure

Small spheres are sodium
and large spheres chlorine
ions. Each ion is surrounded
by, and attracted to, six ions
of opposite sign (Wyckoff)

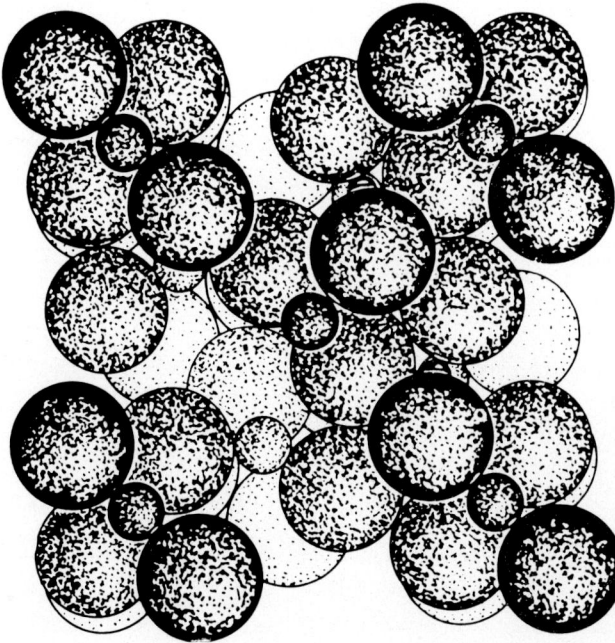

Fig. 2
Model of sillimanite
crystal $Al_2O_3SiO_2$
which is closely similar
to that of mullite
$3Al_2O_3 2SiO_2$

Small spheres are metal, the
larger oxygen, atoms. (Like
the cubic sodium chloride
structure, this orthorhombic
structure can be continued
indefinitely in any direction
without any large holes being
left.) Each atom is surrounded
by, and bound to, several
others; e.g. each silicon is
surrounded by four oxygens
(Wyckoff)

Fig. 3
Models of hydrocarbon chain molecules

(a) Two lengths of —CH_2— chain in straight configuration lying approximately as in crystalline regions of paraffin wax or polyethylene, but slightly separated for clarity. The dark central core represents the spine of carbon atoms, the light hemispheres the hydrogen atoms bound two to each carbon

(b) Portion of chain molecule with short branches, butyl —C_3H_8 and ethyl —C_2H_5

(a)

(b)

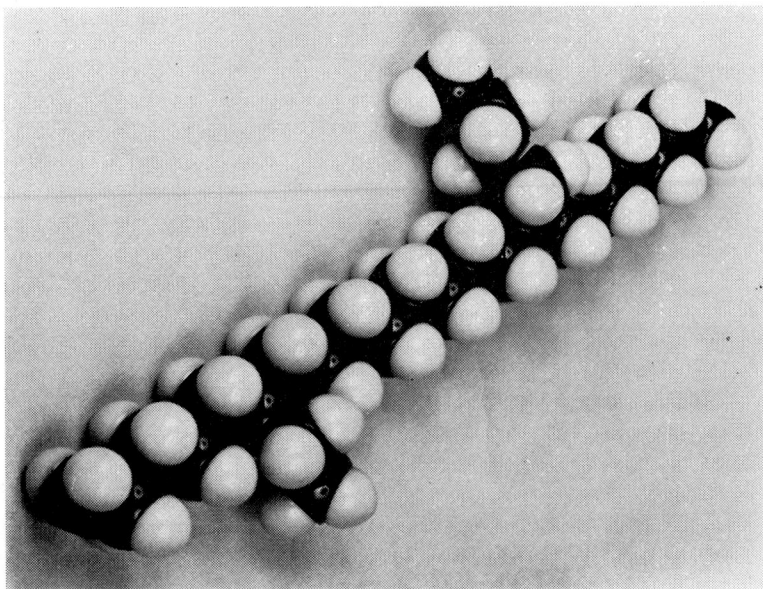

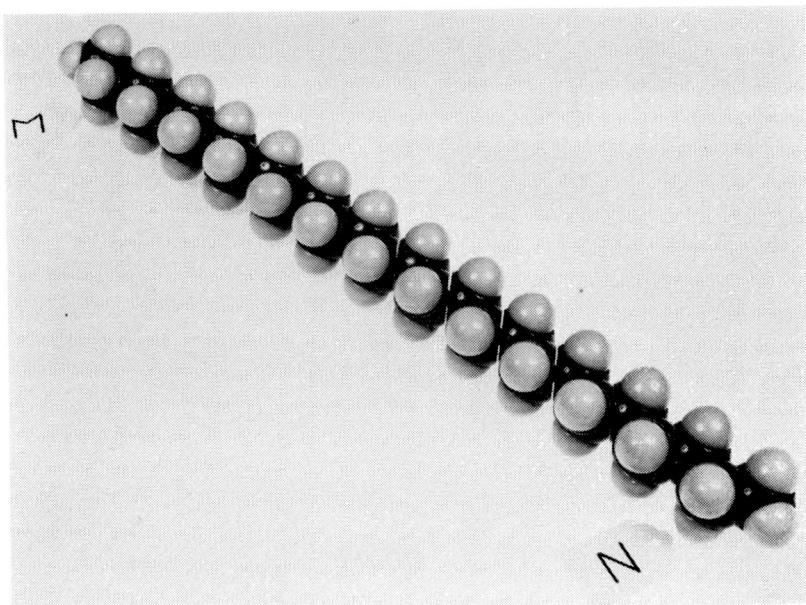

Fig. 9 Models of hydrocarbon chain molecules
(a) 'Straight' conformation
(b) Various random conformations

(a)

(b)

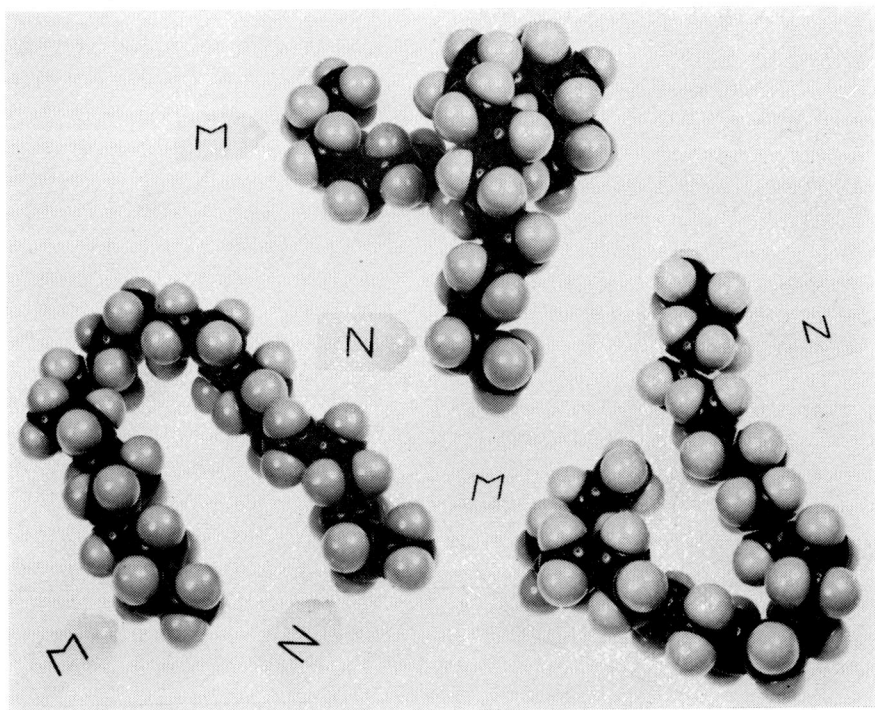

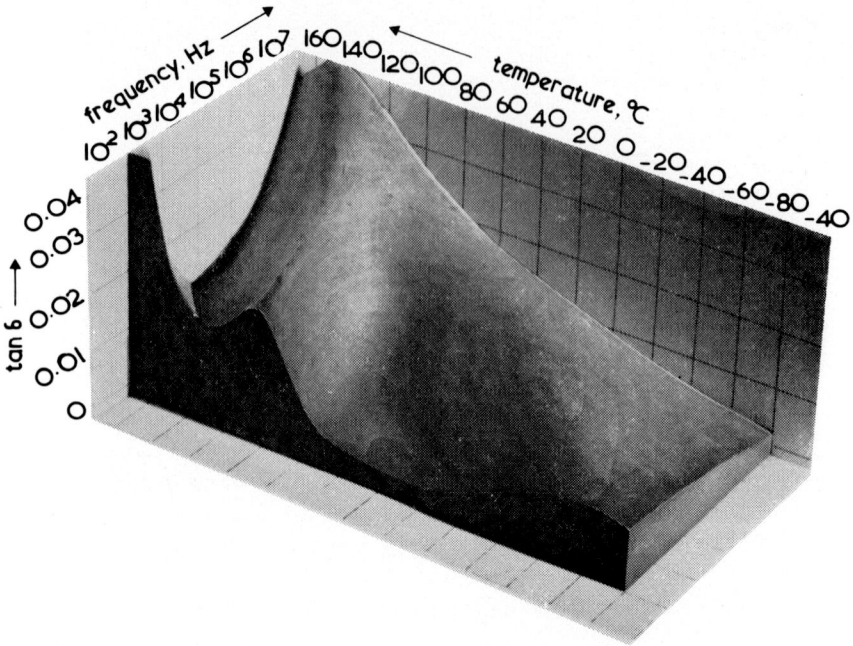

Fig. 20 Relief model of variation of loss tangent with frequency and temperature for poly(ethylene terephthalate)
(Reddish, 1950)

Fig. 21 Relief model of variation of relative permittivity with frequency and temperature for poly(ethylene terephthalate)
(Reddish, 1950)

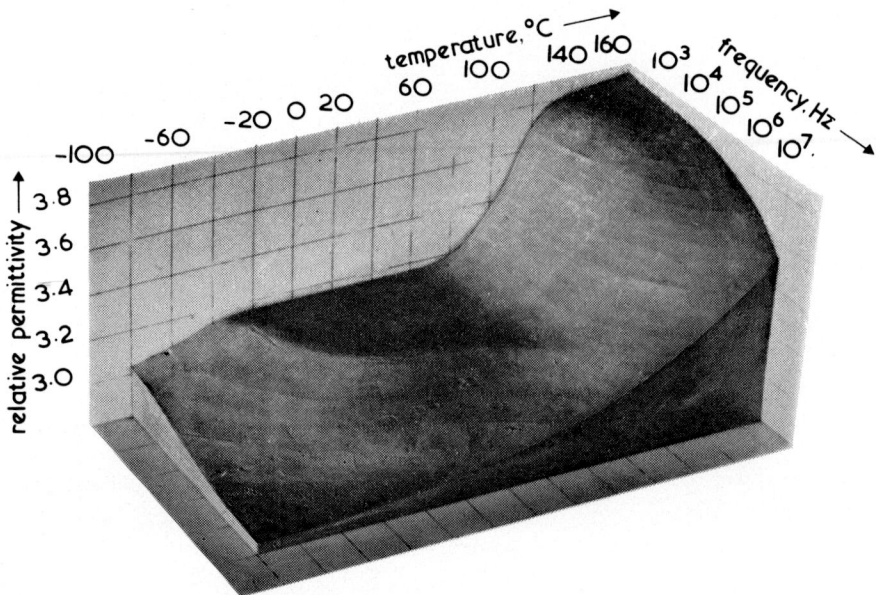

Fig. 27   Tree-like discharge channels in poly(methyl methacrylate)
(Forrest, 1963)

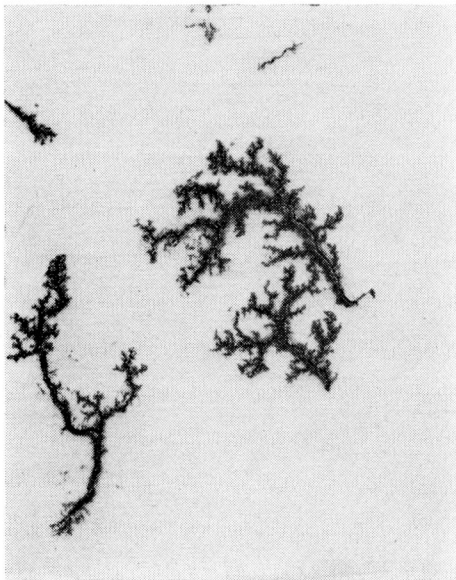

Fig. 30
Tracking produced by a high-voltage
discharge over the surface of a phenol
formaldehyde bonded paper board

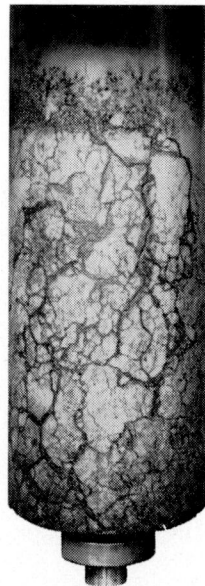

Fig. 31
Tracking on a high voltage bushing
due to water droplets in the oil in
which it was used

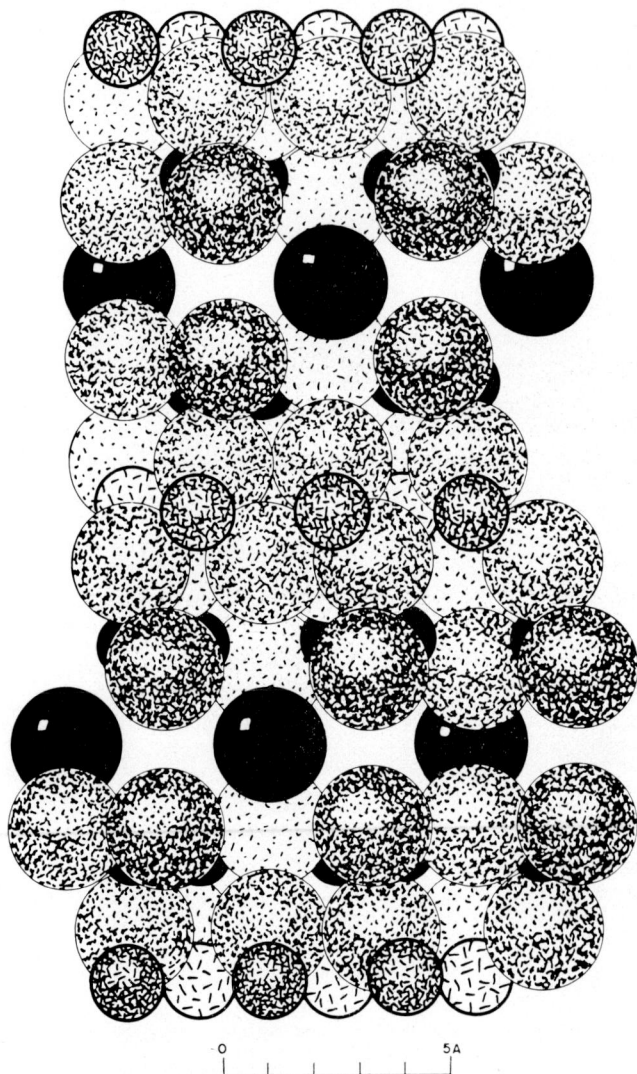

Fig. 36 Model of muscovite mica structure $KAl_3Si_3O_{10}(OH)_2$

The elements are represented as follows:
  potassium: large black spheres
  silicon: small black spheres (largely hidden)
  aluminium: small shaded spheres
  oxygen: large shaded spheres
The hydrogen atoms are not shown. The horizontal planes of potassium atoms are the cleavage planes; layers of alumino-silicate 1 nm thick lie between them (Wyckoff)

Fig. 38 Completing stator of 11 kV vertical-shaft 16-pole water-driven generator
(Courtesy of GEC Ltd.)

Fig. 47 Coil of high voltage generator stator being insulated with mica tape impregnated by B-stage epoxy resin
(Courtesy of GEC Ltd.)

Fig. 54 Early stage of simple dry filament winding of a tube
In wet winding resin is applied to the strands as they leave the guide (Courtesy of GEC Ltd)

a peak corresponding to the transition already mentioned, being higher than $100\,\mu$rad at most frequencies and room temperature, but falling in the microwave region to below $100\,\mu$rad, less than other low loss polymers in this range (Buckingham and Reddish, 1967). It remains uncertain whether the losses as now measured are inherent and due to residual dipole moments in the polypropylene molecule, or partly result from impurities. Figs. 18 and 19 show a comparison between various low loss materials. At $4\cdot2\,\mathrm{K}$ Vincett (1969) found a loss angle of $4\,\mu$rad from very low frequencies to $10\,\mathrm{kHz}$.

8.1.4   Uses of polyethylene, polypropylene and copolymers

The use of polyethylene radio frequency cables has already been mentioned. The dielectric component of attenuation in a cable is proportional to loss tangent and to $\sqrt{(\text{permittivity})}$. For short lengths of cable, and where flexibility and tolerance of rough handling are more important than attenuation, solid polyethylene is used. For minimal attenuation a light structure such as a helically disposed strip, or a cellular (foamed) polyethylene is used to maintain the conductor coaxial with the screen. Irradiated polyolefine sheets are used for microwave stripline printed circuits despite some disadvantage from dimensional instability.

One of the growing uses of olefine polymers, apart from telecommunications cables and components, is for power distribution cables (Hamilton, 1961), where their low loss and absence of 'draining' difficulties and the possibility of dispensing with a watertight sheath, give them advantage over the 'solid' impregnated paper construction. Normal polyethylene is not used because of stress-cracking, nor is simple polypropylene because of its low temperature brittleness. Cross-linked polyethylene is the commonest material, ethylene-propylene, ethylene-butene and ethylene-propylene-diene are also used, ethylene-propylene rubber especially being widely used for moderate voltage flexible cables. Reynolds and Edwards (1970) discuss the uses and limitations of plastic insulation for cables. Barnes *et al.* (1969) point out some of the immediate problems. Ratings up to $33\,\mathrm{kV}$ ($19\,\mathrm{kV}$ to earth) are in common use, and there is considerable experience of higher voltage designs.

The limiting factor as regards high voltage rating of cables is the avoidance of discharge and treeing; it is necessary to provide a smooth electrostatic screen around the stranded conductor and also around the outside of the insulation in contact with the sheath. These may be of semi-conducting (carbon impregnated) lapped

F

paper or a fabric tape, but the most effective systems use a carbon-loaded extruded polymer layer, at least around the conductor, where the stress is greatest (Illers and Kuhmann, 1969) and preferably as the outer screen also (Fujisawa *et al.*, 1968). The two or three layers which result from this construction are best simultaneously extruded by tandem extruders with additional heat for cross-linking applied to the whole immediately afterwards (see references above). Obviously every precaution is needed to exclude gas voids, impurities and gaps between the insulating and the semiconducting layers.

The working stresses employed in single-core solid polyethylene cables from 11–70kV rating have generally been similar to those used for 'solid' impregnated paper, i.e. a mean stress of around 2·5kV/mm, the maximum stress at the conductor (eqn. 7.1) being 40–80% higher according to the conductor diameter (Fujisawa *et al.*, 1968; Barnes, 1966). Higher values have been used experimentally for 70kV cables (Blankenburg, Steinke and Stolpe, 1970) and trial cables for various higher voltages have been made, e.g. up to 225kV with a mean stress in the cross-linked polyethylene insulation of 5·5kV/mm (Charoy and Jocteur, 1971), and a maximum of between 8 and 9kV/mm. None of the plastics used for cables tolerates discharge to the degree that oil- or compound-impregnated paper can.

The requirement for semi-conducting screens to match the insulating plastic and remain in contact with it during thermal cycling has led to several investigations of materials loaded with carbon black; these have conductivities which may be only roughly determined but are usually in the range 10^{-2}–10^{-4} S/m (Bahder and Garcia, 1971; Cesana *et al.*, 1971). In this region of values the conductivity is extremely sensitive to the proportion and type of carbon black and to many other factors. The mechanism of conduction in carbon black itself is not clear in detail; in a dispersion of black in a polymer the passage of electrons between particles which are a finite distance apart is probably involved; no useful basic theory of this process exists because of our vagueness about the mechanism of electron transport through polymers; various empirical theories are employed to try to co-ordinate the facts. The conductivity in an extruded shield appears to be anisotropic, frequency dependent and current dependent. The polymer itself is not necessarily identical with that used for the insulating layer; since the substantial proportion of carbon black needed modifies the mechanical properties, a compensating change in the polymer to obtain overall match may be necessary.

Methods of suppressing discharge, other than semi-conducting screens continuous with the insulation, have been tried; these include pressurising the cable with an electronegative gas (see next paragraph) and filling spaces between conductor and insulation with silicone fluid which does not seriously swell cross-linked polyethylene. Either of these methods adds to the complications of jointing and termination.

Experiments on cables made with lapped high density polyethylene tape suffused with high pressure sulphur hexafluoride have demonstrated the possibility of achieving stresses comparable with those used in high-voltage oil cables (Gibbons *et al.*, 1965; Skipper and McNeal, 1965, Gibbons and Stannett, 1973) but have not yet resulted in a satisfactory cable. One difficulty encountered is the distortion of the tape during thermal cycling.

The possibilities of oil-impregnated polypropylene tape (or oriented polyethylene tape) for cable insulation are being explored (Ball *et al.*, 1972; Edwards *et al.*, 1972). As with capacitors (see below) outgassing and impregnation would be made easier by using alternate layers of paper and plastic. Swelling of the plastic is likely to cause problems which might be eased by the use of paper–plastic laminated tapes. The working stress would probably be similar to that for paper cable of similar voltage rating, but the lower losses would allow a higher current rating of the conductor.

Polyethylene heat-shrinkable sleeves and tapes are used for making joints and terminations in low voltage cables.

Junction and termination boxes for polyethylene and synthetic rubber insulated high voltage cables are now being used, in which no resin or compound is used. A case made of a synthetic rubber such as ethylene-propylene-diene has entry sleeves which are an interference fit with the cable insulation, and contains metal fittings which provide the current-carrying connection mechanically; the joint is effected by removing from the cable the outer protective covering, and baring the conductor, over suitable lengths, then simply pushing the ends into the box. A semiconducting screen connects with the outer screens of the cable ends.

Polyethylene, and to a less extent polypropylene mouldings are now being used for all kinds of cheap electrical components which were previously made of phenolic resin materials, the fall in the cost of fabricated polyethylene and its non-tracking properties have made it attractive for many everyday purposes.

Polyethylene film is rarely made thinner than about $25\,\mu$m, but polypropylene is available as a biaxially oriented film in thicknesses down to $6\,\mu$m, with properties little different from those of the thicker, cast films (Reddish and Tapley, 1970), and suitable for use in two or more thicknesses (to guard against occasional flaws) for power frequency capacitors. (These generally require a greater voltage across the dielectric than do many electronic capacitors.) Polypropylene can be impregnated with the chlorinated diphenyls used for impregnating paper capacitors also used at power frequencies; the polymer swells a little but discharge level and breakdown voltage are improved by the impregnation. There may be some increase of dielectric loss (Kaniskin *et al.*, 1972) but not to the values normal for paper similarly impregnated, (Mercier and McClain, 1968).

It is difficult to achieve complete wetting and impregnation of the tightly wound rolls of foil and polymer, and various means of providing passages for the impregnant have been used. One is simply to emboss or wrinkle the film, but this may lead to instability of the capacitance value. Another is to wind the capacitor with metallised paper and polypropylene film; the metallising acts as electrode, the paper providing a wick for the impregnant. Behn (1968) describes a mineral-oil impregnated capacitor of this type in which a $6\,\mu$m film of polypropylene is stressed at around $70\,\mathrm{V}/\mu$m. The 'burning-out' and isolation of any faults in the dielectric occur as with ordinary paper capacitors, so that a single film of polypropylene can be used. The oil causes some swelling of the polypropylene but does not impair its performance. A loss tangent of 3×10^{-4} at 50 Hz is obtained.

A third method is to form a sandwich of film and paper, using aluminium foil electrodes and trichlordiphenyl impregnant (Curtis, 1970). This is said to improve the ratio of reactive kVA to volume, of a power-factor correction unit, by 20% compared with a normal paper-trichlordiphenyl aluminium foil unit. The loss tangent is generally below 10^{-3} at 50 Hz and of the order of 2×10^{-3} at 10 kHz, falling to lower values during service as is common with chlorinated diphenyl.

From the arguments in Section 4.3.5 it is seen that when the size of a void is only a few micrometres the discharge inception stress becomes very high. It might be expected that voids in impregnated capacitors would be of the same order of size as the dielectric thickness, and that higher inception stresses would be achievable the thinner the film. This is the case, but Krasucki (1967*b*) has given a more detailed explanation, based on the verified assumption that discharge bubbles in the impregnant start at the edge of a foil. The field at the edge is higher in relation to the field in the dielectric the thicker the dielectric film, so the stress in the film to cause discharge inception as the foil edge will be higher the thinner the film; a number of experimental verifications of this are given. Thus the working stress of the film would be expected to be higher for thin films than thick ones. Krasucki (1970) has shown that carefully impregnated polymer film capacitors of polypropylene or polystyrene (two layers each $12 \cdot 5 \mu m$ thick), have discharge stresses in the region of 60–$100 \, V(r.m.s)/\mu m$ and that a substantial proportion of capacitors with these dielectrics stressed at 40–$50 \, V(r.m.s.)/\mu m$ survived several years of life test.

Mat or 'paper' of polypropylene fibres in thicknesses down to $4 \mu m$ may provide a technically attractive, but now expensive, material for impregnated capacitors. Cellulose fibres may be included with the polypropylene to give added strength.

8.2 Polystyrene

The systematic name of the styrene monomer is vinyl benzene

$$\langle \hspace{-0.3em} \bigcirc \hspace{-0.3em} \rangle \; CH{=}CH_2$$

which is made by dehydrogenating ethyl benzene, which in turn is made by various processes from ethylene and benzene. Polymerisation is carried out in the liquid phase, the final stage being carried out in a tower from which the molten polymer

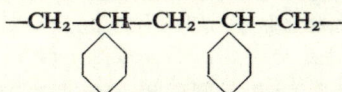

$$-CH_2-CH-CH_2-CH-CH_2-$$

is run off. As normally produced it is actactic and amorphous, of molecular weight $200\,000$ to $300\,000$, it can be made in isotactic form but this has not been produced commercially. It has a glass–rubber transition at about 80°C and a crystal melting point about 240°C. It becomes too soft to withstand mechanical stress around 80–90°C, however. Unmodified, it is brittle at room temperature with a tendency to crazing; this can be overcome by plasticising or copolymerisation, with a lowering of softening temperature. A small upward extension of the usable temperature range can be got by copolymerising with divinyl benzene

$$CH_2{=}CH\langle \hspace{-0.3em} \bigcirc \hspace{-0.3em} \rangle CH{=}CH_2$$

which produces a light cross-linking ('heat resisting polystyrene'). A tougher material is provided by a copolymer with a small proportion of butadiene or with about 25% of acrylonitrile or other monomers providing high intermolecular forces. One of the major outlets for styrene, its rubbery copolymers with butadiene is discussed in Section 8.14.1.

Unmodified polystyrene is well suited to injection and compression moulding and extrusion, processing temperatures range from 150–280°C. Films up to 150 μm thick are readily produced but are expensive in thicknesses less than 8–10 μm. These films can be biaxially oriented by suitable stretching and heat treatment, with some improvement in mechanical properties. The maximum usable temperature is still about 85°C.

In some applications, such as encapsulation, and in one technique of making cable joints, compounds containing styrene and a peroxide catalyst are used, polymerisation takes place *in situ*.

It discolours and crazes on long exposure to sunlight and cannot be used outdoors. It is prone to oxidation when hot.

It has a very low water absorption, 0·03%, though its properties are more susceptible to deterioration by moisture than those of polyethylene. It is resistant to many acids and alkalis but is attacked by many organic solvents. It is, however, compatible with paraffinic and silicone oils.

The relative permittivity of pure polystyrene is close to 2·55 at all frequencies, and at room temperature its loss angle can be less than 100 μrad at all frequencies up to 10^6 Hz. It tracks fairly readily.

Unmodified polystyrene is flexible only in film form, and apart from mouldings where it is used for its appearance, and mouldings in particular positions in radio frequency equipment where low loss is important, its main electrical uses are for capacitor films and radio frequency cable insulation. For many years it was the only film material from which capacitors could be made; and for many types, notably high stability electronic capacitors and power capacitors, it remains competitive with its later rivals. It has a very high insulation resistance and is used for capacitors where retention of charge for long periods is required. The film can be metallised, but faults do not burn out as with metallised paper because of the copious production of carbon by the aromatic structure.

Polystyrene capacitors are limited to temperatures of 70–75°C. Capacitors for radio frequency circuits are normally operated at stresses of 5–20 V/μm, up to 40 V/μm is possible with silicone oil impregnation. The lower stresses apply to low voltage where the minimum thickness of the film is set by mechanical considerations.

In high-frequency cables it is normally used as strip helically wound round the central conductor, or in the form of cups or beads threaded on to the conductor, these arrangements providing some degree of flexibility. Cellular (foamed) polystyrene may also be used.

Glass reinforced polystyrene has been successfully used as a base for printed circuit boards for high-speed computer circuits; also for microwave stripline circuits (Harper, 1969) which require low permittivity and dimensional stability.

The impact-resisting copolymers with acrylonitrile and similar modifications have temperature limitations similar to those of unmodified polystyrene, naturally they have relatively high dielectric loss.

8.3 Polyparaxylylene

This very specialised material with the structure

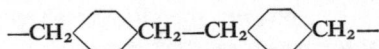

$$-CH_2\hexagon CH_2-CH_2\hexagon CH_2-$$

is produced for low-voltage capacitors by condensation on a cold metal surface of free radicals from the dehydrogenation and pyrolysis of *p*-xylene. A uniform layer of polymer results, which can reputedly be less than 1 μm thick without defects.

The film is easily oxidised in air at about 100°C, but in use for highly compact sealed capacitors it is reported to be stable at up to 200°C. As would be expected from a hydrocarbon the loss tangent is low, of the order of 0·0001; the relative permittivity, 2·63, is similar to that of polystyrene.

This polymer, or a halogen derivative, is sometimes used as a coating for protection of electronic equipment from moisture, or other atmospheric contaminants.

8.4 Polytetrafluorethylene

This unique polymer is produced by catalytic polymerisation of tetrafluorethylene gas, C_2F_4, under pressure, in the presence of an aqueous solution of peroxide catalysts. It is a white powder of the chemical structure

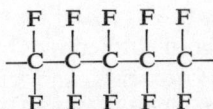

$$\begin{array}{ccccc} F & F & F & F & F \\ | & | & | & | & | \\ -C-C-C-C-C- \\ | & | & | & | & | \\ F & F & F & F & F \end{array}$$

which is believed to be unbranched. Its molecular weight is high, the number-average is thought to be many millions. Its high degree of crystallinity (about 95%), and hence many of its other remarkable properties, are attributed to the stiffness of the molecular chains. The bulk of the two fluorine atoms per carbon atom prevents, at least at ordinary temperatures, the twisting and bending which occurs in polyethylene, and greatly favours the parallel arrangement of straight chains (similar to those of the models of Fig. 3 but with fluorine atoms several times the size of the hydrogen atoms). It undergoes a transition involving a volume change of 1·3% at 19°C, above which some torsional rotation of the chains sets in. It has a crystal melting point at 327°C at which all X-ray evidence of crystalline order disappears, and a volume change of 25% takes place but the polymer does not become fluid; at higher temperatures it decomposes without becoming liquid.

The solid polymer is made by pressing the powder from the polymerisation process under about $14 N/mm^2$ into the required shape, as in ceramic or powder metallurgical techniques, followed by sintering at a temperature such as 380°C, preferably under pressure. It can be extruded slowly on the same principles, a high temperature (400–450°C) and pressure being maintained in the die. The crystal density is 2·30, fabricated products usually have densities of 2·15 upwards.

After forming the material is usually cooled rapidly to avoid the maximum degree of crystallinity, giving better toughness and abrasion resistance.

An aqueous emulsion of p.t.f.e. particles can be produced suitable for dip coating or impregnation. Films down to $6 \mu m$ thick are produced from the emulsion or by skiving (paring with a blade) from solid rod. Unsintered tape for wrapping and sintering *in situ* is available.

The fabricated solid is colourless and waxy, like polyethylene, with a remarkably low coefficient of friction (0·06), tough and resistant to abrasion with a high impact strength, ultimate tensile limit and elongation, also with good endurance to flexing. Despite its intractability at high temperatures it has a pronounced tendency to cold flow at room temperature (followed by some elastic recovery if the stress is removed); this is one of its chief disadvantages. It can be cold drawn at a stress of about $10 N/mm^2$; highly oriented fibres have strengths up to $30 N/mm^2$. Long time tensile properties have been studied by Ogorkiewicz (1970*a*).

It is resistant to all solvents and chemicals except molten sodium (and similar metals) and some chlorinated organic solvents at about 300°C. It cannot therefore be secured by adhesives without special preparation. Its moisture absorption is extremely low, less than 0·01 %. It can be used out of doors.

Its cold flow tendency results in a low heat distortion temperature but this does not imply mechanical strength falling rapidly with increasing temperature. It is chemically stable in air up to 250°C indefinitely and, subject to the mechanical limitations already mentioned, can be used up to that temperature. It does not become brittle at low temperatures and remains usable down to the temperature of liquid helium.

The —CF_2— groups being symmetrical p.t.f.e. is not dipolar, it has a relative permittivity of 2·12, the dielectric loss is very low with a maximum of 250 μrad in the 100–1000 MHz range but with a flat region of less than 30 μrad at frequencies below 1 MHz (see Figs. 18 and 19). The cause of the small loss peak has not been determined, presumably it results from some small concentration of adventitious dipole groups, or from a small unknown dipole moment in the p.t.f.e molecule. At 4·2 K, Vincett (1969) found a loss angle of only 1 μrad at low frequencies up to 10 kHz, Fallou and Bobo find 5 μrad at 50 Hz. It does not track, but is as susceptible as other polymers to discharge.

Electrical uses of p.t.f.e. are generally confined to situations in which a high operating temperature and/or non-inflammability are required, particularly if low loss or resistance to corrosive atmospheres is also necessary. Wire coatings in aircraft and in other special situations, sleeving, slot liners and wire insulation in high temperature machines and magnet coils, capacitors for high ambient temperatures are examples. Flat cables and printed circuits using p.t.f.e. sheet base have been used in high-speed computer and microwave circuits (Harper, 1969); dimensional stability can be improved by glass-fibre reinforcement.

8.5 Tetrafluorethylene-hexafluoropropylene copolymer

This copolymer, usually known as fluorinated ethylenepropylene, has many of the characteristics of polytetrafluorethylene but has a lower molecular weight, a crystal melting point of about 290°C, and melts to a viscous fluid which can be injection, compression or blow moulded. Dispersions are used for wire coating, impregnating windings and impregnating fabrics. Films from 12 μm thickness upwards are available, they can be heat-bonded to other materials, including metals. Mechanical properties are similar to that of p.t.f.e. with a lower, but still excellent, flex life. It has a

F*

low heat distortion temperature, but varnishes and wire coatings are suitable for use up to 180°C. It remains flexible down to 20 K.

Its resistance to solvents, acids and alkalis is similar to that of p.t.f.e. and moisture absorption is less than 0·01 %. Like p.t.f.e. this polymer has a low loss tangent, generally less than 0·0003 except at high frequency, and a relative permittivity of 2·1. It does not track.

Films bonded to metal can be used for printed circuits. Thin films can be used with heat sealing for protecting components and conductors from moisture and solvent.

8.6 Poly(vinyl chloride)

Vinyl chloride monomer, formerly made by addition of HCl to acetylene, C_2H_2, is now generally made by first reacting HCl with ethylene to form $C_2H_4Cl_2$, then pyrolising to form $CH_2{=}CHCl$. Addition polymerisation, by one of several techniques, generally in suspension, then gives the polymer

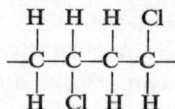

$$\begin{array}{cccc} H & H & H & Cl \\ | & | & | & | \\ -C{-}C{-}C{-}C{-} \\ | & | & | & | \\ H & Cl & H & H \end{array}$$

As normally made it has a molecular weight of 60 000–150 000 and a tendency to be syndiotactic and to have a low degree (order of 10 %) of crystallinity. Some chain branching occurs. It decomposes before its melting point is reached. It is commonly copolymerised with the closely related vinylidene chloride which improves the tensile properties, or with vinyl acetate for increased ease of fabrication. Copolymers with diethyl fumarate or diethyl maleate are also used. [Poly(vinylidene chloride) is not used as a homopolymer because it decomposes too readily on heating, poly(vinyl acetate) softens at a temperature close to room temperature.]

Poly(vinyl chloride) is compatible with a large number of plasticisers and other polymers may be mixed with it; mechanical properties of the very numerous formulations range from the rigid horny material familiar in rainwater gutters to the highly flexible sheets equally familiar in furnishings and handbags. Both rigid and flexible types can be fabricated by any of the usual techniques, processing temperatures being 150–200°C. They can be joined by welding. The rigid materials usually contain a few per cent of epoxy resin, have good abrasion resistance and are unaffected by aliphatic liquids, but swell or dissolve in several aromatic and chlorinated solvents. Soft plasticised materials have a lower tensile strength than the rigid ones, with elongation of 200% or more: common plasticisers are tributyl phosphate, dibutyl phthalate and similar esters. Often polymers such as nitrile rubber or other elastomers are mixed with it. Most compositions exhibit at some temperature a delayed partial recovery after deformation, similar to a lightly cross-linked 'lazy' rubber. This has been attributed to the existence of occasional crystalline regions acting as weak cross-links. The temperature at which this occurs depends on the degree of plasticisation. Some long-time mechanical properties of p.v.c. compositions are recorded by Ogorkiewicz (1970a).

Stabilisers such as long-chain fatty acid salts of calcium, barium, lead etc., or other types of metallo-organic compounds, are incorporated to minimise decomposition at high temperature or by exposure to sunlight.

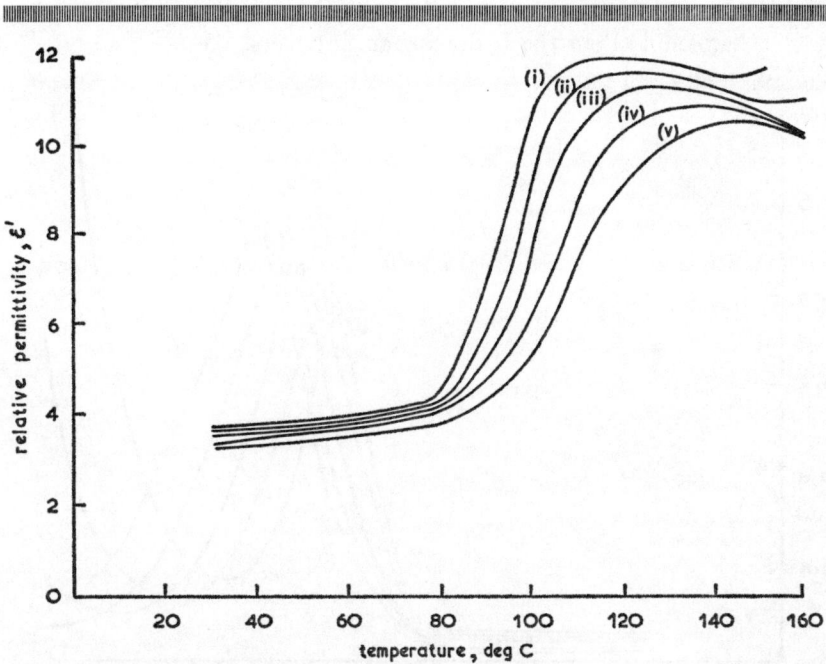

Fig. 41 Variation of relative permittivity ε' of poly(vinyl chloride) with temperature and frequency
(Reddish, 1966)

(i) at 32 Hz	(iv) at 5620 Hz
(ii) at 178 Hz	(v) at 31 600 Hz
(iii) at 1000 Hz	

Suitable compositions can be moulded but the majority of electrical applications are for extruded wire coatings, sheets, films, tubes etc. Plastisols and organisols (suspensions of polymer in liquid plasticiser with or without organic solvents) are used for producing plastic coatings by, for example, dipping, followed by heating to 150–160°C.

The maximum temperature of use is often limited by mechanical considerations, but chemical instability restricts long-term temperatures to 100–110°C (the appendix to IEC 85 puts it in class Y— 90°C). High temperatures are possible for short times but prolonged heating above 115°C leads to some production of hydrochloric acid with the possibility of corroding conductors etc. The highest temperature for continuous use of p.v.c. cable insulation is 70–85°C, depending on formulation; there is no lower limit to temperature of use after installation but cables cannot be manipulated at low temperatures, 0 to −20°C, again depending on the formulation. Many p.v.c. materials are self-extinguishing, p.v.c. generally is less flammable than the hydrocarbon polymers.

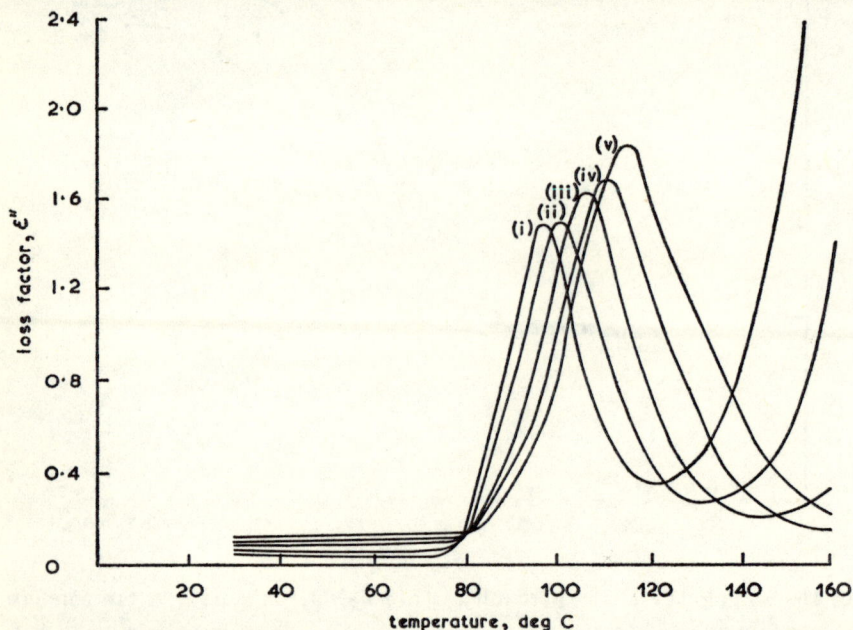

Fig. 42   Variation of loss factor $\varepsilon'' = \varepsilon' \tan \delta$ of poly(vinyl chloride) with temperature and frequency
(Reddish, 1966)
Key as for Fig. 41

Water absorption by rigid p.v.c. may be 0·5–1%, suitable mineral fillers with surface treatments can keep this low; co-polymers with vinyl acetate may absorb as little as 0·1%, highly plasticised compositions up to 2%.

As would be expected from a substance with such a high concentration of groups of high dipole moment, permittivity and loss can be high and there is a very broad spread of relaxation times. Relative permittivities from 3·5 to 10 or 12 and loss tangents from 0·006 to 0·2 are often quoted. Figs. 41 and 42 show how the relative permittivity and loss factor of an unplasticised p.v.c. vary with temperature and frequency. The effect of plasticising is to reduce the dipole relaxation time and therefore to lower the temperature at which the relaxation time has a particular value. Fig. 43 illustrates this by comparing the 1 kHz curve of Fig. 42 with curves for plasticised p.v.c derived from data in another paper. In the case of the 18·5% dioctyl phthalate sample (curve iii) the maximum permittivity of about 12 would be developed at about 90°C (compared with 140°C in Fig. 41), further plasticising brings

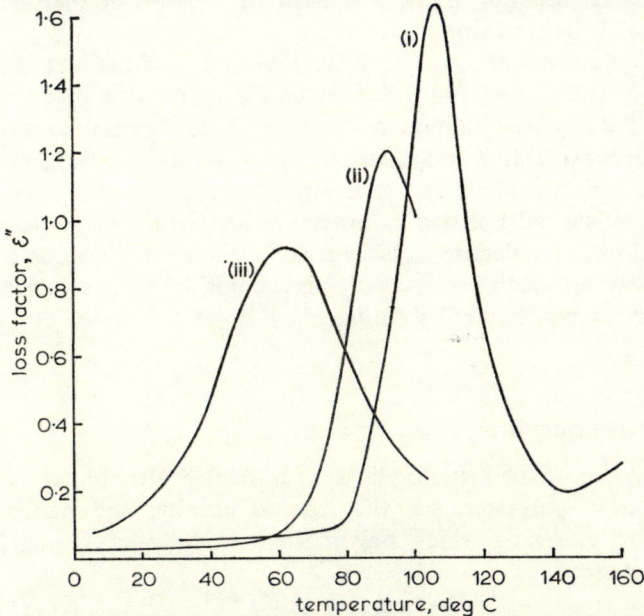

Fig. 43 Variation of loss factor ε'' at 1Hz with temperature
(i) unplasticised p.v.c. (reproduced from Fig. 42)
(ii) plasticised with 4% dioctyl phthalate
(iii) plasticised with 18·5% dioctyl phthalate (curves (ii) and (iii) from Reddish, 1962)

the temperature still lower. Reddish (1966) also demonstrates that as the time scale of the measurement is lengthened to 10^{-4} Hz the rise of permittivity with temperature becomes more and more abrupt, suggesting a transition, discontinuous in temperature but requiring an infinite time to appear, at 74°C (unplasticised p.v.c.), from restricted to unrestricted rotation of the dipoles. The same work also shows that increasing the degree of chlorine substitution above that corresponding to $(C_2H_3Cl)_n$ towards that correspond-ing to $(C_2H_2Cl_2)_n$ *reduces* the change of permittivity and the main dielectric loss peak (as well as moving it to a higher temperature). The second chlorine atom is attached to the same carbon atom as the first, which is then symmetrical and no longer a dipole. Allan and Kuffel (1968) found that at low temperatures the loss angles of a plasticised p.v.c. fell fairly steadily to about 10^{-4} at 4·2 K, and the permittivity remained steady at about 2·65 from 200 K downwards.

Poly(vinyl chloride) has as good an electric strength as most polymers for direct and 50 Hz voltage, but the large losses clearly make it unsuitable for high frequency or high voltage power frequency use; because of the loss itself or because of thermal breakdown. It tracks fairly readily.

It is widely used as insulation for low and medium voltage unsheathed cables, and for jacketing cables of many kinds, on account of its good adhesion and oil resistance. Formulations containing considerable proportions, up to a total of 50% of, plasticisers, mineral fillers and stabilisers are used. The thickness of insulation is usually settled by mechanical rather than electrical considerations, and electric stress is usually below 2 kV/mm. It is not generally practicable to use poly(vinyl chloride) for insulation of cables of higher than 15 kV rating.

8.7 Poly(methyl methacrylate)

This polymer of the methacrylic ester of methyl alcohol can be described as a polyester, but the term is usually, and better, reserved for materials which polymerise via the ester linkage. Methyl methacrylate

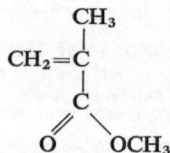

$$CH_2{=}\underset{\underset{\displaystyle O}{\overset{\displaystyle \|}{\underset{\displaystyle C}{|}}}}{\overset{\displaystyle CH_3}{\underset{\displaystyle |}{C}}}{-}OCH_3$$

is unsaturated and polymerises by addition like vinyl chloride to form

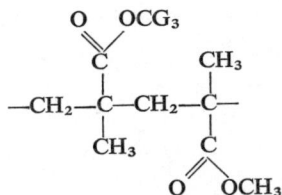

$$
-CH_2-\underset{\underset{\displaystyle CH_3}{|}}{\overset{\overset{\displaystyle \overset{O}{\underset{\displaystyle C}{\diagup\diagdown}} OCG_3}{|}}{C}}-CH_2-\underset{\underset{\displaystyle \underset{O\diagup\diagdown OCH_3}{C}}{|}}{\overset{\overset{\displaystyle CH_3}{|}}{C}}-
$$

Polymerisation is carried out in the liquid phase, in bulk or in suspension. It is 70–75% syndiotactic but because of this incomplete regularity combined with bulky side groups crystallisation does not occur. Molecular weights of up to one million are usual. Completely regular structures can be produced but are not available commercially.

It can be moulded, blow moulded and extruded at temperatures in the range 160–220°C; partly polymerised syrups are available for producing cast shapes, by putting the syrup into the desired mould heated to 60–100°C with a catalyst.

It is a very clear transparent colourless solid with good mechanical strength and impact resistance, but is readily scratched. Its heat distortion temperature is about 90°C, but it is rarely used in electrical applications involving high temperatures. It resists outdoor weathering well and does not track. It resists dilute acids and alkalis, and a number of organic solvents. Water absorption is 0·3–0·4%.

Relative permittivity at room temperature and low frequencies is about 3·6, falling rapidly with rising frequency or falling temperature, loss tangent is around 0·05 falling to 0·01 at radio frequencies and rising to a peak at 40°C and low frequency. Contour maps of these properties are given by Deutsch, Hoff and Reddish (1951), losses are reported over a very wide range of frequencies by Reddish (1962).

Although most of the properties of poly(methyl methacrylate) are acceptable for electrical use it does not have any combination of properties which make it outstandingly suitable for machine, cable or electronic insulation, nor does it readily form films. It is used for cast and moulded parts particularly where appearance is of importance.

8.8 Poly(ethylene terephthalate)

This is a polyester and in principle can be made by the reaction between ethylene glycol $HO-CH_2-CH_2-OH$ and terephthalic acid

$$HO-\underset{O}{\underset{\|}{C}}\left\langle\bigcirc\right\rangle\underset{O}{\underset{\|}{C}}-OH$$

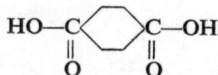

each COOH group of each acid molecule reacting with an OH group of a different glycol molecule or vice versa. In practice this method is not very effective, instead the terephthalic acid (made by oxidising *p*-xylene) is esterified to dimethyl terephthalate, and this is reacted with glycol in an analogous ester interchange reaction:

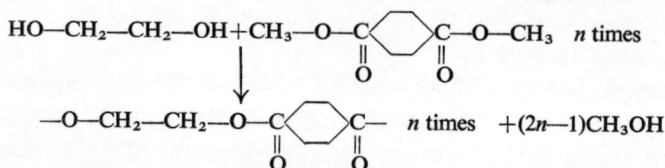

$$HO-CH_2-CH_2-OH+CH_3-O-\underset{O}{\underset{\|}{C}}\left\langle\bigcirc\right\rangle\underset{O}{\underset{\|}{C}}-O-CH_3 \quad n\ \text{times}$$

$$\downarrow$$

$$-O-CH_2-CH_2-O-\underset{O}{\underset{\|}{C}}\left\langle\bigcirc\right\rangle\underset{O}{\underset{\|}{C}}- \quad n\ \text{times} \quad +(2n-1)CH_3OH$$

This polyester is referred to as p.e.t.p. or by one of its trade names such as Terylene (fibre), Melinex, Mylar (film) or simply as 'polyester'*. The number-average molecular weight is normally 20000–30000.

The ability to form remarkably strong films and fibres is attributed to the stiffness of the polymer molecule produced by the *p*-phenylene groups (benzene rings connected in the chain at opposite corners), and the crystallinity and strong interchain forces between the parallel chains. These appear to be the only causes for the great differences in behaviour between this polyester and the linear alkyd polyesters (Section 9.3) formed from glycol and (ordinary) phthalic acid in which the links to the benzene ring are in the neighbouring (*ortho*) positions.

The terephthalate was primarily developed for fibre production; the fibres are spun from the melt, they are considerably stronger than those of cellulose, cellulose acetate or polypropylene. It is readily produced in films by casting, usually followed by stretching to give biaxial orientation; films down to 3–4 μm thick can be produced.

* The word 'polyester' applied to a fibre or film usually implies p.e.t.p.; applied to a solid or bonding resin it usually implies one of the unsaturated cross-linked polyesters (Section 9.5), but it is also used to describe (or disguise) any of the numerous polymers containing ester groupings, the context often leaves some ambiguity

The mechanical behaviour during heating is complex. When quenched from the melt it is amorphous, but it has a glass–rubber transition at about 80°C above which it begins to crystallise, and its mechanical strength diminishes. It can be oriented in one plane by stretching in two directions at about 200°C and maintaining this stretch while it cools. It is then stable unless heated to a still higher temperature than that at which it was heat-set.

Its crystal melting point is 265°C and it retains reasonably good short-time mechanical properties up to 150–175°C, but is not regarded as suitable for long-term use as insulation much above 120°C, in some circumstances it is used at up to 130°C. If heated in the presence of water or in very humid air it will undergo degradation (by hydrolysis) more rapidly than in dry conditions.

The high mechanical strength of the film, and its resistance to failure by repeated flexing are retained down to −20°C and it does not become brittle until it reaches −60°C.

It is subject to attack by strong acids and becomes brittle in contact with phenols and cresols, but is not affected by mineral or silicone oils, or by chlorinated diphenyl. Moisture absorption depends on crystallinity and is of the order of 0·5%. It does not track readily but films are eroded by tracking tests.

Permittivity and dielectric loss of typical terephthalate material are shown in Figs. 17, 20 and 21. At low temperatures Allan and Kuffel (1968) find a peak in tan δ of 0·01–0·02 at 200 K, 50 Hz and at 270 K, 90 kHz, falling to about 2×10^{-4} at 4·2 K. Permittivity drops to about 2·55 at 100 K and below.

The film is very widely used for capacitor manufacture; its strength and tear resistance, the ease with which it can be metallised, and above all the thinness of films which can be produced and successfully handled balance, for many types of capacitors, the disadvanatages of an appreciable dielectric loss.

Baldwin and Hamilton (1971) state that terephthalate foil capacitors employ working stresses up to 32 V/μm at temperatures up to 85°C and half this up to 125°C. With metallised films of 1–2Ω/square, faults are self clearing, and these authors suggest that higher stresses may be practicable with film of the order of 6–10 μm thick. (Note that with a 10 μm film 32 V/μm corresponds to 320 V, at which discharge can barely take place). Rice (1971) has described a construction consisting of terephthalate film with a cellulose acetate coat 2·5 μm thick on each side, metallised electrodes being deposited on the acetate whose function is to ensure the clearance of faults by the gas it generates on heating. For the very specialised purpose of pulsed energy storage capacitors of finite life it is suggested that a 25 μm film can be stressed to 2500 V.

As tape or sleeving it is used for many purposes such as binders and heat barriers in multicore communication cables, magnetic recording tape, flat cables, transformer winding insulation and various applications where high strength and heat shrinkage are required. It is the basis of many adhesive insulating tapes. In the power field film and sheet are used for slot insulation in small motors, where exposure to solvents or refrigerants may occur or where strength and toughness are needed with a somewhat better heat stability than cellulose materials provide. (Cross-linked polyester wire coatings are mentioned in Section 9.3.)

A plastic, polybutylene terephthalate is described by Lovenguth (1972) with a tensile strength of 50 N/mm^2 unreinforced and 120 N/mm^2 reinforced. Heat deflection temperatures (18·5 kgf/cm^2) are given as 70°C and 220°C, respectively, and moisture absorption about 0·1 %. Its electrical properties and thermal limitations are not dissimilar from those of p.e.t.p. It is processed at temperatures from 230 to 270°C and suggested uses are for small mouldings such as radio potentiometer cases, connectors, bobbins etc.

8.9 Polycarbonate

This polymer could be regarded as the polyester formed by reacting bisphenol A (diphenylol propane)

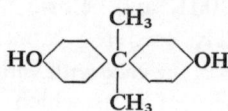

with carbonic acid $\begin{smallmatrix} HO-C-OH \\ \| \\ O \end{smallmatrix}$, but in practice carbonyl chloride (phosgene) $O{=}CCl_2$ is used instead of carbonic acid (hydrochloric acid being eliminated), or diphenyl carbonate is used in an ester exchange similar to that used for preparing p.e.t.p. giving the structure

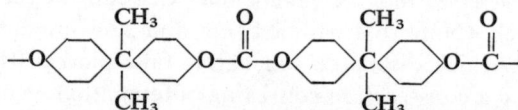

It is a clear, transparent material, has a fairly strong tendency to crystallise and can be obtained in either amorphous or crystalline form, generally a mixture of both, of molecular weight 30000 for

normal processing, and up to 90000 for solvent casting. Though rigid and form stable it has a high impact strength as well as a high tensile strength. The impact strength diminishes considerably at temperatures below $-10°C$ due to a rubber–glass transition, it follows that it is likely to behave in a relatively brittle manner towards very high speed impacts at more normal temperatures.

It can be injection or compression moulded or extruded into sheets etc. at processing temperatures from 250 to 330°C. It can be cold-worked to some extent by rolling and the film can be oriented by cold drawing to give a unidirectional strength of about $200 N/mm^2$. Mouldings tend to have rather high residual stresses. It lends itself to the production of films as thin as $2 \mu m$.

Its crystal melting point is 268°C; it is mechanically usable up to 130°C, or higher in the case of crystalline film, and can be used for long periods at 120°C without deterioration.

It absorbs only about 0·3 % of moisture on immersion but it is not suitable for electrical applications in damp conditions as it tracks readily, and may undergo hydrolysis and degradation.

It is resistant to mineral oil but partly soluble in aromatic fluids and other organic solvents and is attacked by alkalis and strong acids.

The relative permittivity is near to three at room temperature, falling by a few per cent. between 50 Hz and 1 MHz; the loss

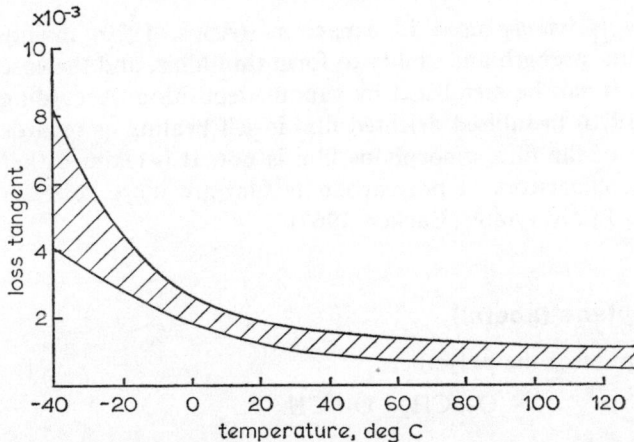

Fig. 44 Representative variation of loss tangent with temperature of poly-carbonate at 1 kHz

tangent is around 0·001 at low frequencies rising to some ten times this in the megacycle range. Figs. 44 and 45 give typical loss angle curves against temperature and frequency.

In moulded form it is generally used where high impact strength and transparency are required. Although readily available, its use is limited, doubtless by its cost.

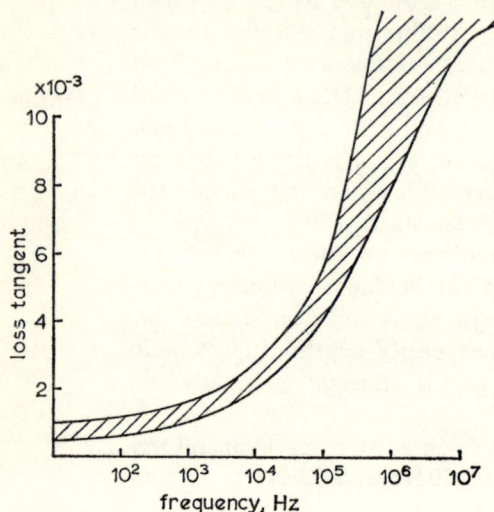

Fig. 45 Representative variation of loss tangent with frequency of polycarbonate at room temperature

The film is widely used in capacitors (Apps, 1970) mainly because of its strength and ability to form thin films, and the ease with which it can be metallised by vapour deposition. According to Reese (1970) metallised oriented film is self healing as regards breakdown of the film, amorphous film is not. It is claimed that low voltage capacitors of polycarbonate film are more compact than those of polystyrene (Hacker, 1967).

8.10 Polyoxymethylene (acetal)

This simplest of the polyethers

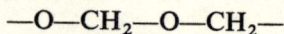

$$-O-CH_2-O-CH_2-$$

is made by addition polymerisation of very pure formaldehyde $H_2C=O$ by an anionic catalyst in an organic solvent, and is usually called acetal resin [not to be confused with poly(vinyl acetal), see Section 8.11]. It is normally about 75% crystalline, with a molecular weight of 30000 and a crystal melting point of 180°C; it has a high impact strength, abrasion resistance and yield stress, and is generally insoluble in organic solvents at room temperature. A number of commercial resins, including copolymers with ethylene, are produced for manufacture of extruded and compression moulded parts, processing temperatures being around 200°C. Mechanical properties at room temperature are rather similar to those of metals and acetals are often used in place of corrosion resisting metals in mass-produced components. Long-term mechanical properties have been investigated by Ogorkiewicz (1970*a*).

Electrically their properties are acceptable for many purposes, dielectric strength is normal, relative permittivity is 3·7–3·8 and loss angle in the region of 0·001–0·004. They are resistant to tracking but burn fairly readily.

They are rarely used in applications where their dielectric properties matter, but bobbins, brush holders, switch parts and components where the need for insulation with good mechanical strength justifies the cost are often made from them.

8.11 Poly(vinyl formal), poly(vinyl acetal), poly(vinyl butyral)

This class of polymers, known generally as the polyvinyl acetals [although poly(vinyl acetal) itself is not common,] must not be confused with 'acetal' (Section 8.10). They are made by condensation of polyvinyl alcohol

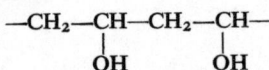

$$-CH_2-CH-CH_2-CH-$$
$$ | |$$
$$ OH OH$$

with the appropriate aldehyde, eliminating the alcohol groups giving, with formaldehyde, poly(vinyl formal):

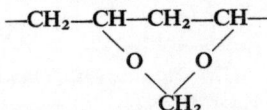

$$-CH_2-CH-CH_2-CH-$$
$$\backslash \diagup$$
$$O O$$
$$\backslash \diagup$$
$$CH_2$$

and with butyraldehyde poly(vinyl butryal)

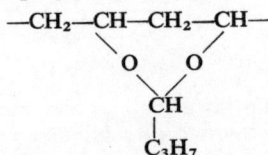

$$-CH_2-CH-CH_2-CH-$$
$$\backslash \diagup$$
$$O O$$
$$\backslash \diagup$$
$$CH$$
$$|$$
$$C_3H_7$$

(Polyvinyl alcohol is made by hydrolising polyvinyl acetate, which, in turn, is obtained by reacting acetic acid with acetylene and polymerising.) They are amorphous, of molecular weight 50000–500000.

Poly(vinyl formal) is a tough abrasion-resistant material which can be cast, but is rarely used for any electrical purposes but wire coating ('enamel') for which purpose it is modified with phenolic resins, and often with isocyanate and butylated melamine resins. These compositions are very suitable for coating wire for magnet and motor windings, having a good combination of abrasion resistance and smoothness or 'windability'. They have water absorption of 0·5–1 % and some formulations are prone to solvent crazing. They are not suitable for use above 105°C and are usually protected by an impregnating varnish which must be chosen to be compatible* with the coating.

The relative permittivity of poly(vinyl formal) compositions usually lies between three and four, increasing rapidly with temperature; the loss tangent at 50 Hz, of the order of 0·005, also increases rapidly with temperature, but the dielectric strength remains entirely adequate for wire coating.

Poly(vinyl butyral) is mostly used plasticised, as an adhesive in non-electrical applications, especially for laminating safety glass. It is also used as an 'overcoat' on wire coated with other resins to produce a winding which will be self-bonding on heating, as an alternative to impregnation.

8.12 Polyamides, polyimides and combinations

8.12.1 Alkyl polyamides

The linear polymers known collectively as the nylons, despite their importance as fibres, and to some extent as solid plastics in the general field, are relatively unimportant as electrical insulation. In structure they are analogous to linear polyesters: the basic reaction is that of a dibasic acid, such as the 6-carbon adipic acid $HOOC(CH_2)_4COOH$ with a diamine such as hexamethylene diamine $H_2NCH_2(CH_2)_4CH_2NH_2$, eliminating water to form a polymer linked by amide, —CO—NH—, groups:

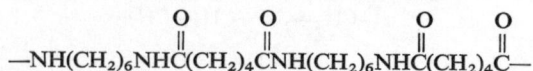

$$-NH(CH_2)_6NH\overset{\overset{\displaystyle O}{\|}}{C}(CH_2)_4\overset{\overset{\displaystyle O}{\|}}{C}NH(CH_2)_6NH\overset{\overset{\displaystyle O}{\|}}{C}(CH_2)_4\overset{\overset{\displaystyle O}{\|}}{C}-$$

* The compatibility of wire coatings and varnishes is usually assessed in the first place by the 'twisted wire' test and more fully by the motorette test (see Section 6.2.5.4. and 6.2.5.5). The Insulation/Circuits Encyclopedia, has a table of compatible combinations.

The system of nomenclature gives an immediate indication of composition, the first figure being the number of carbon atoms in the diamine, and the second the number in the acid. Thus nylon 66 is the poly(hexamethylene adipamide) just mentioned, nylon 610 is similar but with sebacic acid $HOOC(CH_2)_8COOH$ instead of adipic. Other linear polyamides can be made from the reaction of an amino-acid with itself, such as the 6-carbon amino-caproic acid $H_2N(CH_2)_5COOH$ forming a repeating chain of five —CH_2 groups linked by amide groups; this is poly(ω-caproic acid) or polycaprolactam, nylon 6.

While they make extremely strong and abrasion-resistant fibres with a high melting point (above 200°C) and retain mechanical strength up to temperatures of 125–150°C they are not recommended for electrical applications above 105°C even when protected by other resins. Their electrical properties are severely affected by moisture, even within the normal range of humidity of the atmosphere, and they degrade by hydrolysis in the presence of humidity at high temperatures.

Because of its surface properties nylon 66 is, however, used for magnet wire coatings or for overcoats on other coatings less vulnerable to moisture. Its high resistance to solvents permits it to be combined with most impregnating varnishes other than silicones and makes it useful as sheathing for protection of industrial and aircraft cable insulation from oil or petrol. It is also used for lacings and as a plastic filler and reinforcement for thermoplastic boards.

8.12.2 Aromatic polyamides

A typical aromatic polyamide results from reacting the symmetrical diamino diphenyl ether

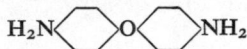

$$H_2N\langle\ \rangle O\langle\ \rangle NH_2$$

with terephthalic acid $HOOC\langle\ \rangle COOH$ to form a linear polyamide similarly to the alkyl ones considered in Section 8.12.1. The molecular chains of aromatic groups are much stiffer than their alkyl counterparts giving substances which do not melt and are much more stable to oxidation and moisture. These are often called 'high temperature polyamides' or just 'polyamides' (the alkyl ones being called 'nylons'), sometimes with a good deal of ambiguity.

Like the nylons the aromatic polyamides form fibres which are available as yarns and fabrics, but are more commonly used as papers or mats which can be impregnated with suitable varnish to provide high temperature insulation.

This type of polymer does not melt but degrades rapidly at temperatures above 370°C. Its mechanical properties and resistance to abrasion are comparable to those of nylon but unaffected by moisture; it is similarly resistant to hydrocarbon solvents and many acids and alkalis. It can be used continuously at temperatures at least up to 180°C and reputedly for long periods at 220°C. It is self-extinguishing.

The papers and mats are available in compressed heat-bonded form and also in a blend with exfoliated mica aiming to increase the discharge resistance and tracking resistance.

Aromatic polyamide fabrics in a variety of combinations, woven with glass and in combination with polyester and polyimide films are made. Polyamide mat bonded to both sides of a terephthalate film has been suggested for 155°C motor slot insulation.

The loss tangent is naturally high. The high proportion of aromatic groups and hydrogen bonding confers resistance to high-energy radiation.

Uses are obviously for high-temperature insulation in dry type transformers, slot liners for class F motors, and hot spots where temperatures of 180–220°C are possible. As varnishes do not wet it readily it is difficult to impregnate and suitable only for low voltages.

8.12.3   Aromatic polyimides

An extension of the use of hydrogen bonding and stiffening by aromatic groups to confer high temperature stability leads to the structure

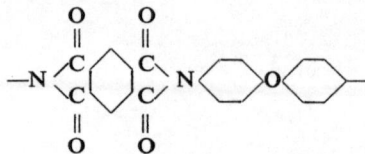

obtained by the reaction of diamino-diphenyl ether (cf aromatic polyamides, Section 8.12.2) with pyromellitic anhydride

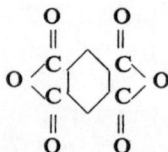

an intermediate stage being the formation of a polyamide-acid in which only half of each anhydride group has reacted with the diamine. This intermediate or prepolymer can be cast into films or compression moulded before the final cure in which the insoluble infusible polyimide is produced. Other similar reactants are also used.

Polyimide is made in the form of film of about $25\,\mu$m thick and above, sheet, varnish prepolymer, wire coating and glass cloth laminates. It retains 70% of its strength at 200°C and is suitable for long term use at 250°C. It survives a few hours at 400°C, above 800°C it chars. It does not burn. The film retains flexibility at liquid helium temperature.

It is resistant to all organic solvents and to acids, but attacked by strong alkalis. It is not, however, recommended for use under oil when paper is also present. Moisture absorption is appreciable, various figures from 1 to 3% and higher are quoted. Water is produced during the conversion of the prepolymer to the final polyimide and must be driven off during curing.

Among organic materials it is one of the most resistant to high energy radiation.

Relative permittivity at 25°C and 1 kHz is 3·5 and loss tangent 0·003 or higher. The loss tangent does not increase significantly up to 200°C but rises rapidly at higher temperatures. The film tracks fairly easily.

Polymer chains with alternations of ester and imide linkage (polyester-imide), or of amide and imide linkages (polyamide-imide) having properties (and prices) intermediate between the two types are available, providing easier processing, greater flexibility and less production of water during final processing, but lower thermal stability than the polyimide homopoloymer. The properties of materials from various sources are summarised by Almouli and Bruins (1968).

The uses of this type of polymer are mostly those where ability to withstand high temperature combined with toughness and flexibility justifies the cost. Polyimide wire coatings, and impregnating varnishes for use up to 220°C in transformers or machines are examples, these being unaffected by all solvents, but the wire coating is, naturally, not solderable.

The ester-imides are suggested for class H machines and may be suitable for higher temperatures in some circumstances. Polyimide film can be used as interlayer transformer insulation, or for motor slot liner material in traction and other high temperature motors where the high strength at these temperatures may allow thinner materials to be used, partly offsetting the high cost. Heat-sealable films or wire coatings, having an outer layer of fluorinated ethylene propylene (sealing at 300–400°C) are available for taping conductors etc. and for self-consolidating windings, respectively. Glass fabric laminate or film can be metallised or copper clad and used for printed circuits, preferably sealed by bonding with a fluorinated ethylene propylene layer. The mechanical strength and low loss of the film make it an acceptable material for capacitor manufacture.

8.13 Polysulphones

The condensation of bisphenol A (see Section 8.9) with 4–4 dichlordiphenylsulphone $Cl\langle\ \rangle S\langle\ \rangle Cl$ gives the polymer

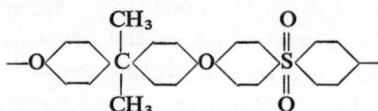

which with similar polymers has become available in sheet and film form with good mechanical and thermal properties and processing temperatures of 300–350°C.

Commercial polysulphone is a transparent flexible material, reputedly usable at temperatures from −50 to 150°C.

It is resistant to acids and alkalis but soluble in many organic solvents. Its moisture absorption is of the order of 0.2%; it is less susceptible to hydrolysis than polyester films.

Relative permittivity lies between 3·1 and 3·0 over the range from 50 Hz to 1 MHz at room temperature (DeMatos and Hutzler, 1971). Loss tangent at room temperature and 1 kHz is about 0·001 rising to 0·003 at 1 MHz, rising also at low temperatures and 1 kHz to 0·004 at −50 °C. Values obtained at higher temperatures are conflicting.

This is among the expensive polymers and seems mostly to have been considered for capacitors required to perform at high temperatures.

8.14 Elastomers

The general nature of elastomers—linear amorphous polymers with occasional cross-links produced by vulcanisation or as part of the polymerisation process and without which the rubber would be a viscous fluid—has been described in Section 2.4. It follows from consideration of this structure that the temperature range of rubber-like behaviour is limited, at least with hydrocarbon polymers, because the thermodynamic temperature at which oxidation embrittlement, or some other form of degradation, sets in is likely to be 25–50 % higher than that at which the interchange forces between chains are sufficiently overcome or the molecule becomes sufficiently flexible to allow the movements required. It follows also that the distinction between an elastomer and a flexible plastic is somewhat arbitrary; many flexible plastics, e.g. p.v.c., display some degree of recovery after deformation and release, all elastomers recover at a finite rate and recover incompletely in a finite time.

The theory of elastomeric behaviour outlined in Section 2.4 predicts that elastic modulus will increase as the thermodynamic temperature, because the higher the temperature the greater the tendency to restore randomness. While this is true of natural rubber over a certain temperature range, the modulus of many other rubbers decreases with increasing temperature. The theory, of course, assumes a fixed number of cross-links but if weak interchain bonds are acting as cross-links these will becomes progressively less effective at high temperatures, and the net effect may be a reduction of modulus.

All rubbers are affected in some degree by some solvents but some resist the action of particular classes of solvents well, usually when the molecular types of the polymer and the solvent are dissimilar.

Elastomers tend to be vulnerable to oxidation, probably because a relatively small degree of oxidation is sufficient to increase the number of interchain bonds to the point of embrittlement, and because any residual double bonds provide points for oxygen attack. For the latter reason natural rubber and SBR are more prone to oxidation than other synthetic rubbers.

Many rubbers develop their most desirable properties only when reinforced with carbon black or mineral fillers, and modified with extenders (plasticisers) or blended with other rubbers; so that only a rough indication of properties related to particular types can be given. It is, indeed, fairly common to find statements about the relative qualities of rubbers which are, superficially at least, mutually contradictory; tables of properties are not very useful.

The majority of electrical uses are for wire and cable insulation (Reynolds and Edwards, 1970), moulded plugs and connectors especially for outdoor use and where rough usage may occur, and for hand-held and portable equipment. Natural rubber still finds use in flexible cords, but for many purposes it has been superseded by cheaper materials and those more stable to oxidation.

Ethylene-propylene-copolymers and ethylene-propylene-diene vulcanisable polymers are now widely used elastomers for installation and distribution cable insulation; these have been considered in Section 8.1 and its subsections.

There is a great variety of elastomeric polymers, many of them containing halogens or sulphur to give good solvent or oxidation resistance and increase the usable temperature range. These include chlorosulphonated polyethylene, ethylene-vinyl acetate copolymer, various fluoro-hydrocarbons, fluorosilicone, polychloroprene, acrylonitrile-butadiene (NBR) copolymers, polysulphides, polyurethanes and acrylate co-polymers. Many of these are used as jackets for protection of other insulating polymers, or may themselves be used as low voltage insulation. Polychloroprene, for example, is used on trailing cables because of its flame resistance, oil resistance, and abrasion resistance. Broadly speaking, resistance to solvents or high temperatures is achieved at the expense of one of the desirable mechanical as well as the electrical properties.

These are described in detail by Roff and Scott (1971) and by Blow (1971); their properties are discussed in the context of electrical uses by Clark (1962).

Here we will deal individually with only three elastomers, styrene-butadiene copolymer, polyisobutylene (butyl rubber) and silicone elastomers. Comparisons of these three, including also polychloroprene, are given by Green and Verne (1960).

The permittivity and power factors of elastomers are principally dependent on their dipolar content, and to some extent on the presence of carbon black and other fillers, when used. The materials containing halogens, sulphur and nitrile groups have permittivities of seven and over, and high losses; the hydrocarbon elastomers without vulcanisation or oxidation have permittivities in the region of 2–3 and loss tangents from 0·002 to 0·005. Relaxation times for dipoles in elastomers are usually in the milli- and microsecond region, increasing vulcanisation of natural rubber has generally been found to increase the loss tangent and produce an increasingly broad, flattened loss peak, the effects on synthetic rubbers are different for different rubbers.

8.14.1   Poly(butadiene styrene) copolymer

Generally known as SBR (formerly GR–S, and in Germany Buna-S) this is the synthetic rubber most widely used for general purposes. It is polymerised in suspension, or in solution, the highest molecular weights and for most purposes the best materials are polymerised at temperatures around 5°C (cold rubber) and contain about 20% of styrene.

Butadiene CH_2=CH—CH=CH_2 and styrene (see Section 8.2) copolymerise to give molecules of many possible configurations, such as:

$$
\begin{array}{c}
CH_2 \\
\parallel \\
CH \\
\mid \\
=CH\!-\!CH_2\!-\!CH\!-\!CH_2\!-\!CH_2\!-\!CH\!=\!CH\!-\!CH_2\!-\!CH\!-\!CH_2\!- \\
\end{array}
$$

The double bonds in the side-chains provide the necessary conditions for vulcanisation by sulphur or other means (though the mechanism of this is complex).

SBR is compounded like natural rubber, extended with hydrocarbon oils and greatly used in blends with other rubbers. Carbon black is used as reinforcement. It has a density of about 0·94 g/ml, mechanical properties are generally a little inferior to those of natural rubber, particularly at elevated temperatures, but it is somewhat more resistant to embrittlement by oxidation in air. The tear resistance is poor but abrasion resistance is good except at high temperatures or after ageing. In cables the continuous conductor temperature is limited to 60°C. At low temperatures it remains flexible down to −40°C. It burns with a smoky flame.

It is not resistant to mineral oils or to solvents in general. Its water absorption and transmission are very low.

Payne (1959) measured the loss and permittivity of SBR over a range of frequency from 50–10^6 Hz and temperature from −50 to 50°C for various degrees of vulcanisation. He used a process of curve fitting to deduce the relationship between relaxation time and temperature, and the trend of loss tangent over a wide range of frequencies. He found that increasing the degree of vulcanisation increases the peak value of loss tangent from about 0·04 for unvulcanised material to 0·14 for 6% sulphur, above which no further increase in peak height or width occurs, but the relaxation times at constant temperature are increased by a factor of about 10^4 in changing from unvulcanised material to 10% sulphur. Power frequency permittivity at room temperature changes from about 2·7 to 3·0 with this change of vulcanisation but goes up to 3·4 at higher temperatures for the more vulcanised material and down as low as 2·3 at low temperatures for the unvulcanised. Loss tangent at room temperature and power frequency is about 0·005 for the unvulcanised and 0·05 for the highly vulcanised material.

Payne also measured the dynamic mechanical behaviour over the same range of variables. He found a mechanical tan δ peak of about two; this did not increase in height with increasing vulcanisation but became a little wider and flatter, moving to lower frequencies in the same way as the dielectric loss peak. The relationship between temperature and change of relaxation time was the same for both electrical and mechanical behaviour.

These electrical and mechanical properties are typical rather than reproducible from one sample to another.

8.14.2 Polyisobutylene—butyl rubber

Isobutylene $\begin{array}{c} CH_2 \\ \| \\ CH_3-C-CH_3 \end{array}$ forms a linear polymer which cannot be vulcanised, and it is normally copolymerised with 2–4% of isoprene $\begin{array}{c} CH_2=C-CH=CH_2 \\ | \\ CH_3 \end{array}$ in the presence of aluminium tri-chloride at −95°C. This copolymer is known as butyl rubber, formerly as GR-I

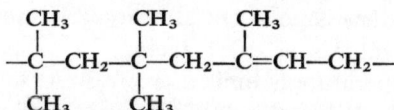

$$-\overset{\overset{\displaystyle CH_3}{|}}{\underset{\underset{\displaystyle CH_3}{|}}{C}}-CH_2-\overset{\overset{\displaystyle CH_3}{|}}{\underset{\underset{\displaystyle CH_3}{|}}{C}}-CH_2-\overset{\overset{\displaystyle CH_3}{|}}{C}=CH-CH_2-$$

It can be vulcanised to a degree depending on the proportion of isoprene, special vulcanising agents being used. Having no double bonds after vulcanisation it does not embrittle by oxidation. It has a density of 0·91 g/ml, moderate tensile strength, good tear resistance at elevated temperatures and good abrasion resistance. It swells in the majority of solvents but not in vegetable oils. At normal temperatures it is sluggish, only at about 90°C does its behaviour resemble that of natural rubber at room temperature.

Since it does not readily oxidise it tends to soften rather than embrittle on heating, and can be used for periods at about 150°C. Conductor temperatures in butyl rubber cables are usually limited to 85 or 90°C continuously, 105°C in emergency overload and 200°C on short circuit.

Relative permittivity is in the range 2·35 to 2·45 (the higher values relate to vulcanised material), loss tangent variation at room temperature is shown in Fig. 46.

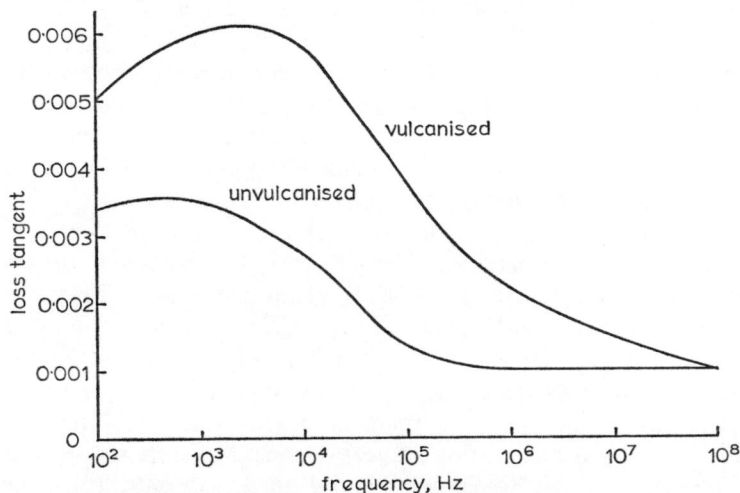

Fig. 46 Variation of loss tangent with frequency of an unvulcanised and a vulcanised butyl rubber at room temperature
From data of Greene and Verne, 1960

Normal butyl rubber is not invulnerable to tracking, but formulations are now produced which do not track in outdoor use at high voltage.

8.14.3 Silicone (polysiloxane) elastomers

The linear polysiloxanes, exemplified by poly(dimethyl siloxane), are usually obtained starting from a methyl halide and silicon which are reacted at high temperature to give, for example, dimethyl chlorsilane. The halide is hydrolised and the resulting hydroxyl compounds undergo a condensation reaction

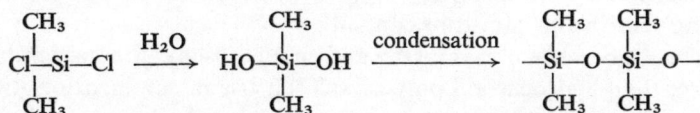

$$\underset{\overset{|}{CH_3}}{\overset{\overset{CH_3}{|}}{Cl-Si-Cl}} \quad \xrightarrow{H_2O} \quad \underset{\overset{|}{CH_3}}{\overset{\overset{CH_3}{|}}{HO-Si-OH}} \quad \xrightarrow{condensation} \quad \underset{\overset{|}{CH_3}\ \ \overset{|}{CH_3}}{\overset{\overset{CH_3}{|}\ \ \overset{CH_3}{|}}{-Si-O-Si-O-}}$$

Other groups besides —CH_3 may be included, phenyl-C_6H_5 being common, and other methods of synthesis are in use.

The cross-linking necessary to give elastomeric behaviour may be produced by peroxide curing at 150–200°C, by including some vinyl groups in place of the saturated groups and reacting them, or by providing additional OH groups attached to some silicon atoms which take part in the condensation reaction to produce interchain Si—O—Si links. The latter method is used for room-temperature curing rubbers, either in the form of a two component mix, or a single paste, one component of which is hydrolised by atmospheric moisture when exposed to it, with the production of acetic acid or, in formulations for electrical purposes, methanol.

Silicone rubbers are soft, of density slightly less than water, mechanically somewhat weaker and more easily torn than the commoner elastomers, with poor abrasion resistance and a tendency to permanent set. Tensile strengths are 6–11 N/mm² except for the cold cured materials which are weaker. They are usually reinforced with silica or other mineral fillers to improve their mechanical properties. This increases density to 1·1–1·3 g/ml. Most silicone rubbers swell in solvents, including some mineral oils, without undergoing permanent degradation. Swelling can be lessened by incorporating a fluorosilicone. Moisture absorption is low, but may be higher for mineral filled materials. They are resistant to sunlight and can be used out of doors.

These rubbers are outstanding for their endurance to high temperatures, withstanding 250°C for long periods, but prone to degradation in the presence of water vapour. When employed for cables conductor temperatures of 180°C may be used continuously. When heated in a sealed enclosure they are said to degrade slowly above 175°C. Materials suitable for use at −70 to −100°C are made, but normal grades remain flexible only to about −50°C. The elastomers burn leaving an insulating ash of silica; when pyrolised in the absence of air they produce methanol.

The permittivity of pure dimethyl siloxane polymer is about three with loss tangent from 0·0005 to 0·0001 over the range from power frequency to kilomegahertz. However the pure rubber is rarely usable and the incorporation of inorganic metal oxides to improve its mechanical properties may increase permittivity to the region of 7–10 and loss tangent to 0·001, 0·01 or higher, both being sensitive to moisture content.

The rubbers are non-tracking and endure discharge considerably better than hydrocarbon polymers. Their use in combination with mica for high voltage machine insulation has been considered (Rembold, 1964; Galpern and Brown, 1971).

They are used as extruded coatings for high temperature cables, e.g. in aircraft. For high temperature machines and transformers, tapes, sheets and sleeves, unsupported or with glass fabric reinforcement, in thicknesses from about 100 μm upwards are used for covering conductors and busbars. Where self adhesion is required the tape or sheet is made from partly cured rubber, or uncured rubber coated on to cured rubber, and adhesion is obtained by heating.

A very low loss form of silastomer known as dielectric gel, supplied in the form of two mobile liquids which gel on mixing and heating, may be used for encapsulation of delicate components, the material being so soft that little stress can be imposed; it is also used for cast insulation or for impregnation. Elastomers vulcanisable at room temperature (having high loss angles) may be used similarly, and for encapsulating motor windings etc. Particular silicone rubber formulations have also been used for coating epoxy-glass components for use outdoors, special care to ensure adhesion being necessary.

The single-component moisture-catalysed rubber (mentioned above) is used for filling spaces, and for dip and spray coating for moisture sealing. General accounts of silicone elastomers have been given by Davis and Rees (1965) and by Kookootsedes and Dexter (1968).

G

Chapter 9

Cross-linked synthetic polymers

It is usual to state or imply that linear polymers are thermoplastic, i.e. become viscous liquids on heating and solidify on cooling, an alternation which can be repeated indefinitely; cross-linked polymers on the contrary are taken to be thermosetting, the heat curing establishes the cross-links simultaneously with the final stage of polymerisation, the resin becomes solid while still hot and can never be resoftened but remains insoluble and infusible, decomposing at high temperatures without melting. These useful generalisations can occasionally cause confusion: linear polymers with rigid molecular chains such as polytetrafluorethylene and the polyimides are insoluble and decompose without liquefying but are normally counted as thermoplastics because they do not become more rigid when heated for long periods and their molecules remain linear; cross-linked polymers such as epoxides, on the other hand, are regarded as thermosetting, but some formulations may still be partly soluble in strong solvents and may be softened by heating. Shellac is an intermediate case where the softening temperature may be raised by heating. A further disparity is that certain cross-linked resins can be cured to rigid and relatively insoluble solids without heating. Remembering that elastomers are essentially linear polymers lightly cross-linked it is not difficult to see why moderately cross-linked materials may soften on heating and swell in solvents. From a practical viewpoint the labels thermoplastic and thermosetting are usually appropriate and are widely used but may occasionally seem misleading.

We have seen that linear polymers are formed from monomer molecules each of which reacts with two other molecules (whether by addition as with ethylene or by condensation as with polyesters), such monomers are called bifunctional and can only produce linear chains. Cross-linking obviously requires a proportion, at least, of monomers which are polyfunctional—able to react with three or more other groups—and it is obvious that unless all the monomer molecules are identical (which is not often the case) a variety of different structures can appear. The structural chemistry of a thermosetting resin is therefore more complex than that of a thermoplastic and often cannot be precisely described.

Consequently the technology of cross-linked polymers is more empirical than that of linear ones.

It is common to speak of three stages in the preparation of a thermosetting resin: the A-stage is a thermoplastic soluble resin, capable of being cross-linked but suitable for use as a varnish or for polymerising with other synthetic resins or natural drying oils; the B-stage is produced by further heating and reaction and is a cross-linked solid, soluble only in powerful solvents, hard at room temperature but softened by heating, suitable for moulding powders and preimpregnated tapes and fabrics; the C-stage is the final completely infusible insoluble material produced by further heating of the B-stage material when formed into the shape finally required, it is typified by a fully cured phenolic moulding.

Several of the cross-linked polymers are now widely used for glass fabric laminates and other glass-fibre reinforced materials. The production, use and properties of these laminates is a subject on which there are a number of text books; a recent one on materials and manufacture is by Lubin (1969); Parkyn (1970) discusses in detail the mechanical properties and the design problems resulting from them; both of these carry references to works on various individual materials.

The families of thermosetting resins made from formaldehyde and the phenols (phenol, cresol etc.), and those made from formaldehyde and the amides or amines (urea, melamine etc) have many variants, these families are often referred to as the phenoplasts and the aminoplasts respectively.

9.1   Phenolic resins

These resins might qualify for inclusion in Chapter 7 since they have been in use for longer than modern designs of electrical equipment. They were developed rather empirically and, while the main patterns of polymerisation are now well known, there are many variants and many modifications with other polymers, particularly with the natural drying oils, which remain empirical.

The phenol formaldehydes are typically made by reacting

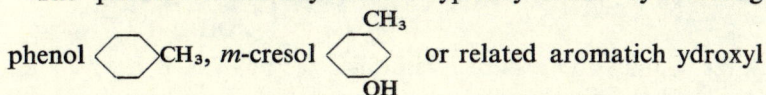

phenol $\langle\bigcirc\rangle$CH₃, *m*-cresol $\underset{OH}{\overset{CH_3}{\langle\bigcirc\rangle}}$ or related aromatich ydroxyl compounds with formaldehyde $HC\!\!\begin{smallmatrix}O\\\\H\end{smallmatrix}$ or furfuraldehyde (a heterocyclic aldehyde obtained from oats). The first step in the reaction with formaldehyde is its addition in the *ortho* (neighbouring) or *para* (opposite) position (relative to the hydroxyl group) to give a methylol group ⬡OH ; this may be followed by CH_2OH further additions in the remaining *o-* or *p-*positions if sufficient formaldehyde is present. There are then two routes to the A-stage material (see, for instance, Brydson, 1969).

The single stage route, often used for production of laminating resins, consists in heating the formaldehyde and phenol in molecular ratio about $1\cdot25:1$, with an alkaline catalyst. The methylol groups condense with hydrogens on the aromatic rings of other molecules to form chains of aromatic rings linked by $-CH_2-$ groups

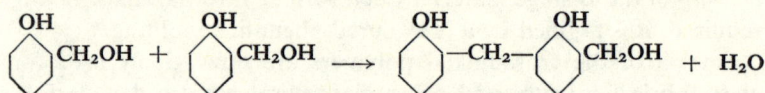

$$OH\bigcirc CH_2OH + OH\bigcirc CH_2OH \longrightarrow OH\bigcirc -CH_2-\bigcirc OH\ CH_2OH + H_2O$$

and so on, to produce the thermoplastic resin known as 'resole'. Alternatively, the methylol groups react with each other to form an ether link which may then decompose to give the same result as above. Since there is excess of formaldehyde, some phenol molecules acquire two methylol groups and can become cross links. The reaction is stopped while the resole is still fusible and soluble, cooled and the product ground and mixed with filler or dissolved in alcohol. Paper or cloth to be laminated is impregnated with this solution and dried. It can then be stored ready for forming into sheets by hot pressing, or into tubes or bushings by winding on mandrels under hot rolls. The final cure may be completed, if necessary, by oven baking.

The two-stage route begins with heating the phenol with a slightly less than equimolecular proportion of formaldehyde so that there is little chance of a phenol acquiring more than one methylol group and only the linear polymer is formed by a process similar to the one described above. This thermoplastic resin is called 'novolak' and usually consists of about 10 aromatic rings per molecule:

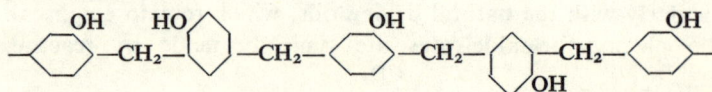

$$-\overset{OH}{\bigcirc}-CH_2-\overset{HO}{\bigcirc}-CH_2-\overset{OH}{\bigcirc}-CH_2-\bigcirc_{OH}-CH_2-\overset{OH}{\bigcirc}-$$

After cooling and grinding with filler the novolak is heated with hexamethylene tetramine ('hexamine') which reacts with moisture to produce formaldehyde and an A or B stage resin is produced, usually on hot rolls. Again the reaction is stopped short of full curing, ground and prepared as moulding powder. The final cure of phenolics is usually carried out at 140–170°C.

Moulding resins are nearly always made with fillers which may be mineral, cellulose fibre, asbestos fibre, glass fibre, wood flour etc. Densities of mouldings range from 1·25 to 2 g/ml according to type of filler used. Mechanical properties may be roughly summarised by saying that tensile strengths are in the range 50–70 N/mm^2 with less than 1 % elongation at break; hardness (Rockwell) is around M100, elasticity modulus 5000–20 000 N/mm^2 impact strength only moderate but highly dependent on the type of filler and on modifications of the resin. Moisture absorption generally lies between 0·1 and 1 % depending on filler. Most compositions are slow burning or self-extinguishing depending on the filler or reinforcement. The limiting temperature for continuous use for electrical purposes is usually between 105 and 130°C. The lower limit applies to varnished cellulose cloths and papers, the higher to modified phenolics when used as a bond for inorganic materials. Still higher limits are quoted for asbestos board. In applications where embrittlement would be a serious drawback the lower limit is usually applied.

Fully cured resins are resistant to nearly all solvents, acids and alkalis, but this does not necessarily apply to laminates.

The cheapest laminates are paper-based (the phrase 'synthetic resin-bonded paper' or s.r.b.p. usually refers to phenolic bond) but all kinds of mats and fabrics are used. The properties indicated above generally apply to laminates as well as to moulding powders but tensile strengths may range up to 140 N/mm^2, while water absorption for cellulose-base laminates with low resin content may be several per cent. Laminates with a high resin content on the other hand are usable in tropical conditions.

The polymerisation processes already indicated begin in aqueous solution and involve water-producing reactions throughout. Thorough curing and baking to convert hydroxyl groups and remove water are necessary if good dielectric properties are required; low dielectric loss cannot be achieved.

A well-cured resin has relative permittivity of about 4·5 and a loss tangent of the order of 0·01, both increasing with increasing temperature and decreasing with increasing frequency; with some fillers permittivity becomes 5–7 and with cellulose filler or laminate the loss tangent is often between 0·01 and 0·05. With hygroscopic fillers which are not fully dried, both loss and permittivity may be very high.

Most phenolic materials track in some degree, those containing cellulose as filler or laminate base do so very readily. This can be partly ameliorated by coating with glyptal or other track-resistant varnish.

Phenolic mouldings and laminates must be processed with care if they are to be used in equipment, such as a sensitive relay, which is vulnerable to small degrees of corrosion. Traces of the original reactants, especially phenols, in sealed equipment may give rise to malfunction after several years. Phenolics also permit silver migration if there is a direct voltage between silver or silver-plated components in contact with the resin. A filament of silver grows out from the anode, and may ultimately produce a short circuit.

Phenolic resins modified with natural drying oils or alkyds give flexible impregnating varnishes which are compatible with many of the materials used for magnet wire coatings (see table of compatibilities in Insulation/Circuits Encyclopaedia, Section B12) and can be used to impregnate motor windings for use at temperatures of 105–130°C depending on the composition used. Hard phenolics are used in high-speed motors and in sealed motors where exposure to refrigerants or other solvents can occur, but are normally restricted to a continuous temperature of 105°C. Phenolics can also be modified with unsaturated polyester or epoxy resins, the latter imparting strong adhesion to metals. Moulding materials prepared with a proportion of furfural resin allow a long moulding period but cure rapidly at high temperature. A modification with nitrile rubber gives high impact resistance, but the use of glass fibre reinforcement is more effective in this respect.

Laminated, resin bonded paper boards, tubes etc. are used in nearly every field of electrical engineering, more especially where exposure to moisture is unlikely and tracking is not expected to be a problem; we will not attempt to list their applications. Many grades suitable for different purposes are defined in BS 1137, BS 2076, BS 2572, BS 3953, DIN 7735 and NEMA publication LI–1, similar tables are given by Birks (1960), Clark (1962) and Oburger (1957). Cotton and asbestos laminates are also covered by BS 2572, BS 2966 and by the German and US standards just mentioned. Glass-phenolic laminates are less frequently used because the high pressure necessary during curing of phenolics tends to damage the glass fibres, these are covered by NEMA LI–1, BS 3953 and DIN 7735. Copper-clad phenolics for printed circuits are dealt with in BS 3888 but the more recent BS 4584 accords with the international standard IEC 249; NEMA LI-1 and DIN 40802 also cover copper-clad boards. A recent summary of laminate types and uses is given by Donelly (1971).

Synthetic resin bonded paper bushings, generally with phenolic resin bonds, are widely used, condenser bushings being used in the higher ranges of voltage. Electric stress in these bushings is normally limited to about 1·5 kV/mm, nominal, about half that which is used for oil-impregnated bushings, but in indoor situations, since no outer porcelain case is required, resin bonded bushings are more compact except at the highest voltages. The design of these bushings requires attention to thermal stability, dielectric losses being appreciable; loss tangents are typically in the range 0·005–0·01 at room temperature and 0·02–0·03 at 90°C. Unlike the more highly stressed oil-impregnated paper cables, s.r.b.p. bushings for systems such as 66 and 132 kV will withstand low levels of discharge at working voltage. Jolley (1965) and Marlow (1970) suggest that a discharge level equivalent to just audible hissing is about 100 pC, and that this level is acceptable at working voltage. The ability of bushings to withstand such discharge is probably related to the low average level of stress.

9.2   Urea formaldehyde, melamine formaldehyde and aniline formaldehyde

These are the commonest, and the typical, amino-resins. They are generally similar in properties to the phenolic resins, with some preferable features and a higher price.

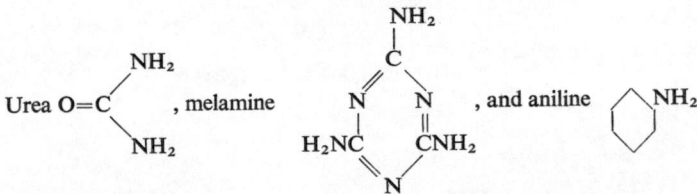

Urea $O=C\begin{smallmatrix}NH_2\\ \\NH_2\end{smallmatrix}$, melamine (triazine structure with NH_2, H_2NC, CNH_2) , and aniline (benzene ring) NH_2

polymerise with formaldehyde somewhat similarly to the phenols in that they accept addition of the formaldehyde molecules, but the addition is usually to the —NH$_2$ group forming $\rangle$NCH$_2$OH groups. These condense with each other to produce methylene links between nitrogen atoms, forming first linear molecules (A-stage), then partly cross-linked (B-stage) and, in the final curing, hard infusible C-stage resins. The partly polymerised resins are aqueous syrups which harden on heating to 90–120°C. Modifications made by reacting the partly polymerised resin with butyl alcohol or other aliphatic alcohols are known as butylated amino resins and are used where compatibility with other, oil soluble, resins is required, or good adhesion to metal surfaces. The various reactions involved in amino-resin formation are described by Martin (1972).

Processing and applications of these resins are similar to the phenolics, but their substantially higher cost restricts their use. Their principal advantages over phenolics are that they are colourless and can be used for white or light-coloured boards (often as a facing on phenolic boards), or for moulded parts of any colour, and that they are much less prone to tracking than phenolics. Views on their relative endurance to high temperatures appear to be conflicting, any advantage in this respect is not sufficient to justify using amino resins. The characteristics of mouldings and laminates are discussed by Birks (1960), Clark (1962) and Oburger (1957); BS 3953 covers two grades of melamine formaldehyde laminate.

Amino resins are also used as adhesives.

9.3 Alkyd resins

Cross-linked polyester resins produced by the straightforward condensation reaction of di- or trihydric alcohols with dibasic acids or anhydrides are usually called alkyds (not polyesters, to avoid confusion with other polyesters). The commonest pair of

reactants are glycerol
$$HO-CH_2-\overset{\overset{\displaystyle OH}{\displaystyle |}}{C}H-CH_2OH$$
and phthalic acid

or its anhydride

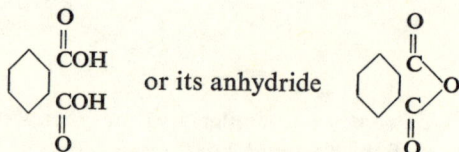

The resins resulting from these are also called glyptals. The first stage is the condensation of one of the —OH groups of the glycerol with one of the acid groups (or half the anhydride group) forming glyceryl monophthalate. A number of these half-esters react successively to form a linear polymer, e.g.

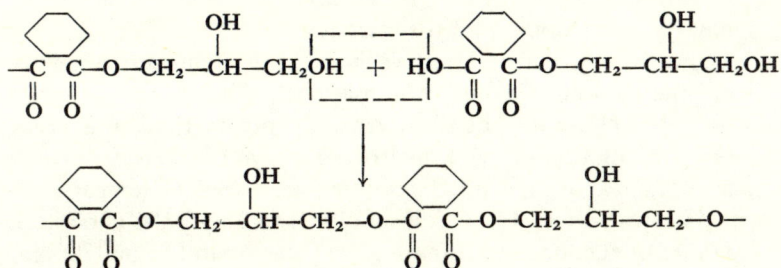

$$\underset{\substack{\|\ \|\\ O\ O}}{-C\ C}-O-CH_2-CH-CH_2OH\ +\ HO-\underset{\substack{\|\ \|\\ O\ O}}{C\ C}-O-CH_2-CH-CH_2OH$$

(with OH groups above the CH, and dashed box around the $-O\ldots HO-$ region)

$$\downarrow$$

$$\underset{\substack{\|\ \|\\ O\ O}}{-C\ C}-O-CH_2-CH-CH_2-O-\underset{\substack{\|\ \|\\ O\ O}}{C\ C}-O-CH_2-CH-CH_2-O-$$

(with OH groups above each CH)

The remaining —OH groups may then react similarly with terminal —COOH groups of other chains or with other OH groups to form —O— (ether) linkages, resulting in a structure of chains of the above type cross-linked together. These reactions are effected at 180–250°C, and finally produce an insoluble infusible resin. To control or prevent this a proportion of monohydric alcohols such as butanol C_4H_9OH, or monobasic acids (generally long-chain acids such as abietic or rosin) is introduced, which has the effect of using up some of the reactive groups in non-polymerising reactions. Alternatively ethylene glycol $HO-CH_2-CH_2-OH$ may be substituted in part for glycerol.

Other constituents employed in producing alkyds are: as alcohols, the higher glycols $HO(CH_2)_nOH$, pentaerythritol $C(CH_2OH)_4$ and other polyhydric alcohols; as acids, adipic acid $HOOC(C_2H_4)_4COOH$, sebacic acid $HOOC(CH_2)_8COOH$, iso-

phthalic acid (benzene ring with COOH, COOH)   maleic acid or its anhydride (maleic anhydride structure: $HC-C$ and $HC-C$ with O atoms)

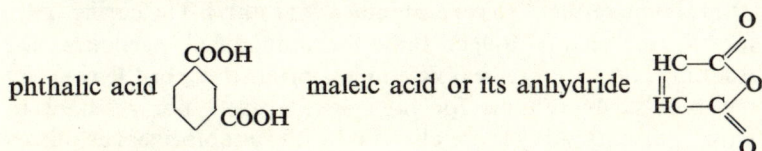

and various long-chain acids.

Terephthalic acid $HOOC$—(ring)—$COOH$, the basis of the linear polyester which is the subject of Section 8.8 is not much used in ordinary alkyd varnishes, cross-linked terephthalate polyesters are however used as magnet wire coatings, referred to again at the end of this section.

The alkyds described above are heat-cured and insoluble in hydrocarbons, but they can be modified with drying oils (such as linseed oil) by adding the fatty acid of the oil to the partly polymerised alkyd, or by reacting the oil with glycerol which is then

G*

used in forming the alkyd. Air-drying alkyds may contain up to 60% of oil, baking varnishes perhaps 30%; conventional paint 'drying' catalysts can be used for curing. Increasing the oil content increases the solubility in hydrocarbons.

Impregnating and coating varnishes are frequently modified with amino resins (giving hard coatings), phenolics or silicones.

Alkyds are usually applied, partially polymerised and cross-linked, in solution in xylol, hydrocarbon-xylol or ketone solvents and after evaporating off the solvent are cured at temperatures ranging from about 130°C (oil modified) to about 180°C (silicone modified). Vacuum impregnation may be desirable but dipping is more generally used; care is needed to drive off the solvents before they are trapped by cured resin.

The relative permittivity of cured alkyds is of the order of 4–5, loss tangent is usually around 0·01–0·05; neither of these properties is important in normal use. The resins are resistant to tracking, more or less so depending on any filler or pigment used, generally they carbonise but do not form a coherent conducting path.

In the non-electrical field they are widely used as paints and protective coatings; in electrical equipment their main fields of use are as coatings on metal parts, on bonded paper insulation as a moisture barrier, as impregnating varnishes for transformers, rotating apparatus and coils generally, and as a bond for built-up mica sheets. As impregnants they are compatible with most wire coating materials except acrylics. The oil- and phenolic-modified resins are suitable for continuous use at 105°C, higher in some circumstances, the silicone combinations generally up to 155°C (these temperatures are general indications only). The curing times for the former are longer than for unmodified phenolics, the resulting resin is more flexible and less prone to embrittlement but not necessarily suitable for high speed rotors nor resistant to strong solvents. Alkyds are also used with suitable fillers as putties for encapsulating small parts, and can be produced as moulding powders. Clark (1962) discusses their uses in some detail.

Polyester wire coatings made with terephthalic acid are rapidly increasing in usage, those made with the reactants already mentioned are suitable for temperatures up to 130°C. Like most polyesters they may be subject to more rapid degradation at very high humidity. Higher temperature coatings are obtained by cross-linking terephthalate with isocyanates, for example a resin made with tris(2 hydroxyethyl) isocyanurate as the polyhydric alcohol. Various combinations of this with aromatic amide or imide over-coats provide coatings which can be used at up to 200°C.

9.4 Unsaturated polyesters

This section deals generally with cross-linked polyesters in which the linear chains are first produced (as in the alkyds) by successive esterifications, but dihydric alcohols and dibasic acids only are used (unlike the alkyds), either the acid or the alcohol having a double bond in its structure (hence the term 'unsaturated polyester') which takes part in the final cross-linking. This final cross-linking is produced in situ with the resin in its final position, by addition polymerisation with some other unsaturated monomer such as styrene which also acts as solvent for the linear polyester during application, impregnation etc. Thus the resin is a liquid of fairly low viscosity during application, but since the 'solvent' styrene is itself incorporated in the final resin it does not have to be driven off before curing, in this sense these polyesters are 'solventless', and therefore used for producing void-free structures.

Commonly they are produced from ethylene glycol $OH-CH_2-CH_2-OH$, propylene glycol or similar dihydric alcohol and the unsaturated maleic anhydride (Section 9.3) or

fumaric acid $\begin{matrix} HOOCH \\ \| \\ HCOOH \end{matrix}$ (maleic acid $\begin{matrix} HCOOH \\ \| \\ HCOOH \end{matrix}$ and fumaric acid

are *cis* and *trans* isomers).

To avoid too great a concentration of double bonds a considerable admixture of the saturated phthalic or adipic acid (Section 9.3) may be used, the higher the proportion of these the softer the resulting resin. The first stage is the reacting of the constituents at 175–200°C to give a linear polymer such as

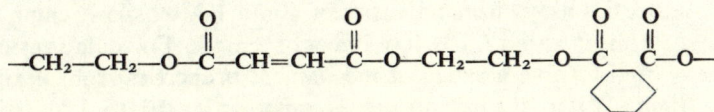

$$-CH_2-CH_2-O-\overset{\overset{\textstyle O}{\|}}{C}-CH=CH-\overset{\overset{\textstyle O}{\|}}{C}-O-CH_2-CH_2-O-\overset{\overset{\textstyle O}{\|}}{C}\bigcirc\overset{\overset{\textstyle O}{\|}}{C}-O-$$

a solid of molecular weight 800–2000.

This polymer is dissolved in styene or some other unsaturated monomer suitable for cross-linking and an inhibitor such as hydroquinone added to enable the resin to be stored. The viscosity can be varied to suit the manner of application by varying the proportion of solvent monomer. When mixed with peroxide initiator and an accelerator it cures at a moderate temperature with the evolution of heat. This heat evolution (exotherm) if not controlled during casting or encapsulation of large volumes may cause strains or cracks in the resin or damage to the object being encapsulated.

The resulting three-dimensional structure cross-linked by styrene can be typified by

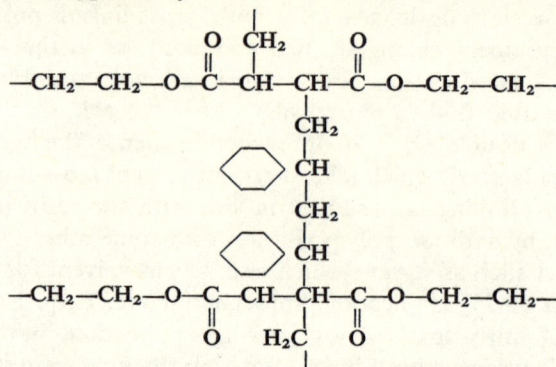

$$
\begin{array}{c}
\text{O} \quad \overset{|}{\underset{}{\text{CH}_2}} \quad \text{O} \\
\parallel \quad | \quad \parallel \\
-\text{CH}_2-\text{CH}_2-\text{O}-\text{C}-\text{CH}-\text{CH}-\text{C}-\text{O}-\text{CH}_2-\text{CH}_2- \\
| \\
\text{CH}_2 \\
| \\
\text{CH} \\
| \\
\text{CH}_2 \\
| \\
\text{CH} \\
| \\
-\text{CH}_2-\text{CH}_2-\text{O}-\text{C}-\text{CH}-\text{CH}-\text{C}-\text{O}-\text{CH}_2-\text{CH}_2- \\
\parallel \quad | \quad \parallel \\
\text{O} \quad \text{H}_2\text{C} \quad \text{O} \\
|
\end{array}
$$

where the number of styrene groups forming the cross-links is variable. Other cross-linking monomers are used such as diallyl phthalate, methyl methacrylate or triallyl cyanurate, the latter conferring increased thermal stability on the resulting polymer. The contraction which takes place during curing is about 9 % if the resin is unfilled, considerably less if filling or reinforcement is used.

Unsaturated polyester is most widely used for simple glass-fibre-reinforced structures, laminates and hand lay-ups built up from wet impregnated glass cloth or mat. These can be cured in the open, or in closed low pressure moulds with heat applied, the catalyst being chosen accordingly. Heat cured resins are used when greater mechanical strength or other improvement in properties, such as dielectric strength, is required. Glass cloth or mica insulation taped on conductors may be vacuum impregnated with the liquid resin and cured.

The resins alone have a density of about 1·3, tensile strength up to 70 N/mm^2 and a fairly low impact strength. Formulations can be altered to give a range of mechanical properties from flexible to rigid. Water absorption values range from 0·1 to 1 %, being generally higher the higher the temperature of curing and the greater the maleic acid content (Clark, 1962). They are not greatly affected by hydrocarbon and non-dipolar chlorinated solvents or by alcohols, but are attacked by many dipolar solvents, by alkalis and by oxidising acids.

Relative permittivity is generally between three and five depending on the proportion and type of cross-linking monomer and on the degree of cure. Loss tangent is generally between 0·01 and 0·05 but can fall outside these limits. Most formulations will track but not easily.

Heat distortion temperatures are low for the soft resins but may be as high as 200°C at 18.5kgf/cm^2 for the hard ones. The highest temperature for continuous use varies with formulation up to 155°C. The resins alone burn slowly.

Filled or reinforced resins are nearly always used; they are among the most convenient and moderate-priced materials for making reinforced structures and are usually the cheapest to fabricate and process (Rubin, 1969; Parkyn, 1970). Glass cloth laminates have densities of $1.5-2\text{g/ml}$, tensile strengths such as 200N/mm^2 and impact strengths of $250-1000 \text{J/m}$. Standard grades are specified by BS 3953 and by NEMA LI–1. Mechanical properties may be adversely affected by prolonged exposure to high humidity or immersion in water (Ogorkiewicz, 1970*b*).

Resins reinforced with glass mat or chopped glass strand have somewhat lower strengths. Dough moulding compounds are made by mixing chopped glass fibre into a putty of the linear polyester and monomer. Filled or reinforced resins are usually self-extinguishing.

Purely electrical uses include impregnating varnish for coils and machine windings (polyester varnishes are compatible with most wire coatings) and for treating glass sleeving. Bare copper in direct contact with uncured polyester inhibits its curing unless special precautions are taken. Some processes for the insulation of high voltage generator stators use a polyester resin bond for mica tape (see Section 7.5.3), these being the first resins of good bonding properties, fluid enough to use as impregnants and not requiring solvent to be driven off, to become avilable. Rogers (1967) describes the development of resins for this purpose, of sufficient flexibility to avoid damage due to thermal expansion or to mechanical distortion. Flexibility in dipolar resins is usually accompanied by high dielectric loss at high temperatures and a compromise is necessary to avoid thermal dielectric breakdown. The resin described by Rogers uses an aromatic dihydric alcohol similar to bisphenol A (see Section 9.5) instead of a glycol, with adipic and fumaric acids in the ratio $1.5:1$. A copper deactivator prevents the inhibition of curing in the layer in contact with the copper.

9.5 Epoxy resins

These resins take their name from the epoxide

$$\underset{\diagdown}{\overset{O}{\diagup}}\text{C}\!-\!\text{CH}\!-$$

group which forms part of the molecule of epichlorhydrin

$$\underset{\text{H}_2\text{C}-\text{CH}-\text{CH}_2\text{Cl}}{\overset{O}{\triangle}}$$ one of the two reactants which are used to make

the resin. The other component most commonly used is diphenylol propane, more often called bisphenol A

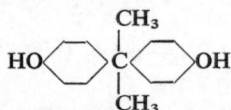

$$\text{HO}\!\!-\!\!\langle\ \rangle\!\!-\!\!\underset{\underset{\text{CH}_3}{|}}{\overset{\overset{\text{CH}_3}{|}}{\text{C}}}\!\!-\!\!\langle\ \rangle\!\!-\!\!\text{OH}$$

When these are heated together in a solution of caustic soda HCl is eliminated between the chlorine of the epichlorhydrin and a hydrogen of one of the hydroxyl groups. At a somewhat higher temperature an hydroxyl group can react with the epoxide group of epichlorhydrin. Both these reactions produce other links and result in a structure of the type

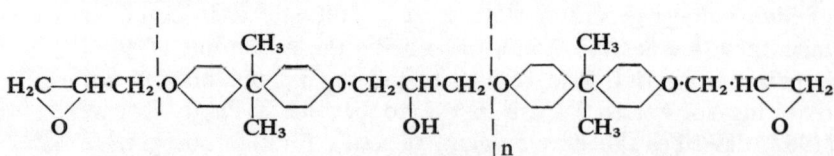

$$\text{H}_2\text{C}\!-\!\text{CH}\!\cdot\!\text{CH}_2\!\cdot\!\text{O}\!\langle\ \rangle\!\underset{\text{CH}_3}{\overset{\text{CH}_3}{\text{C}}}\!\langle\ \rangle\!\text{O}\!\cdot\!\text{CH}_2\!\cdot\!\text{CH}\!\cdot\!\text{CH}_2\!\cdot\!\underset{\underset{\text{OH}}{|}}{\text{O}}\!\langle\ \rangle\!\underset{\text{CH}_3}{\overset{\text{CH}_3}{\text{C}}}\!\langle\ \rangle\!\text{O}\!\cdot\!\text{CH}_2\!\cdot\!\text{HC}\!-\!\text{CH}_2$$

$$|n$$

The residual epoxide groups at either end of the chain result from using a slight excess of epichlorhydrin. This polymerisation may be taken as far as a molecular weight of a few thousand. The degree of polymerisation at this stage is usually described by n, the average number of units of the structure shown between dotted lines above, or by the *epoxide equivalent* which is the average molecular weight of resin per epoxide group, or by the *epoxide value* which is the quotient of dividing the epoxide equivalent into 100. Thus the smallest possible resin molecule, the diglycidyl ether of bisphenol A

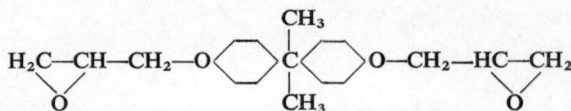

$$\text{H}_2\text{C}\!-\!\text{CH}\!-\!\text{CH}_2\!-\!\text{O}\!\langle\ \rangle\!\underset{\text{CH}_3}{\overset{\text{CH}_3}{\text{C}}}\!\langle\ \rangle\!\text{O}\!-\!\text{CH}_2\!-\!\text{HC}\!-\!\text{CH}_2$$

has an '*n*' of zero, a molecular weight of 340 and a epoxide equivalent of 170. A resin with a uniform '*n*' of 10 would have a molecular weight of 3180 and epoxide equivalent of 1590 but at high molecular weights the epoxide equivalent tends to be greater than half the average molecular weight.

This resin is, of course, thermoplastic, not very soluble in ordinary solvents but dissolved by methyl ethyl ketone, ethoxyethylacetate and similar compounds, which may be diluted with hydrocarbons or alcohols.

A great many different curing agents are used to bring about cross linking of this resin: di- or triamines and amides, the anhydrides of various diacids, and also urea formaldehyde and phenol formaldehyde resins. These react in ways which cannot necessarily be followed in detail (see Roff and Scott, 1970; Potter, 1970) with the terminal epoxide groups, or the hydroxyl groups in the chain, to produce cross-linked resins characteristic of the agent used. The curing releases no water or other product and results in a hard strong heat-resistant resin with very good adhesion to metals and glass as well as to other plastics.

Room temperature hardening, by agents such as aliphatic triamines, is often used; careful control and avoidance of large masses is necessary to prevent the exothermic reaction building up a very high temperature, which may lead to frothing, gas voids, internal stresses and cracks on cooling. The heat distortion temperature of resins hardened at room temperature is generally below 100°C. The best mechanical, electrical and thermal properties are obtained by using the less reactive agents such as aromatic amines, phthalic anhydride or pyromellitic anhydride and hardening at temperatures up to 180°C; a post-cure at temperatures up to 250°C may be recommended. With these methods heat distortion temperatures up to 280°C (at $18 \cdot 5 \, kgf/cm^2$) can be obtained. Single component systems with such hardeners already mixed have a reasonable shelf life at room temperature.

Moulding powders, with mineral or fibre fillers are usually designed to be cured at 100–150°C.

More flexible resins, which are generally desirable for encapsulation and casting with metal components, are obtained by curing with aliphatic amines or by incorporating other polymers. A detailed analysis of the processing and properties associated with various hardeners is given by Dorman (1969).

Castings and low pressure mouldings can be easily made from liquid resins mixed with hardener and filler (if any) at approximately the curing temperature immediately before use, admitted to the mould (which is preferably evacuated if the component is to stand high voltages) and maintained at the curing temperature for the necessary time. Moulds need not be of metal, but may be of wood (if dry), plastic or even rubber. Surfaces must be coated with a release agent, usually a silicone grease.

Laminated fabrics (usually glass fabrics) are made in one of two ways. One is the pre-impregnating resin-rich process, in which the fabric is impregnated liberally with a high temperature curing resin which is solid but flexible at room temperature in its uncured condition. This 'prepreg' can be stored until required; on heating and pressing the resin melts into a uniform mass which then cures. If the uncured resin is not sufficiently flexible (e.g. for taping) it may be softened with solvent which must be removed by vacuum treatment before curing. The other method is to vacuum impregnate a porous structure such as glass cloth or mica tape wound on to a conductor, or a glass filament band or tube, with a low viscosity resin. A variant of the resin-rich process is to apply resin with a brush while building up the glass tape or other structure. Filament winding of tubes is usually done 'wet', i.e. the strands pick up the liquid resin immediately before being wound on the mandrel, but pre-impregnated glass strands can also be used.

Any object to be encapsulated or impregnated must be vacuum dried before introducing the resin if good electrical properties are required.

Casting resins and moulding powders are usually filled with silica, though clay and other inorganic fillers are employed; as much as two parts of filler to one of resin by weight may be used. Chopped glass fibre is used to give impact strength to castings. Fillers must be well dried before mixing. Heat distortion temperatures for cured castings are usually in the neighbourhood 100–110°C (at 18·5 kgf/cm^2).

The higher molecular weight epoxy resins lend themselves to coating of apparatus by fluid bed treatment, the object to be coated being heated at a temperature a little above the melting point of the powder before being suspended over the aerated bed.

The density of the unfilled resin is between 1·1 and 1·3 g/ml, tensile strength 40–80 N/mm^2, modulus of elasticity 2000–10000 N/mm^2 (considerably less for the flexible formulations) and impact strength 20–80 J/m. Filling with silica or other powders does not greatly alter tensile or impact strength but raises the elastic modulus as well as the heat distortion temperature somewhat. Mould shrinkage of cast resins is between 2 and 4%, reduced to about $\frac{1}{2}$% by quartz filler.

Glass fabric laminates with epoxy resins are the strongest available, with tensile strengths 200–400 N/mm^2 and impact strengths 300–1000 J/m. Properties of standard grades are defined in BS 3953 and in NEMA LI–1 and DIN 7735, and are discussed by Donelly (1971). Detailed accounts are given by Lubin (1969) and Parkyn (1970).

Heat distortion temperatures range from 80 to 230°C (at $18·5 kgf/cm^2$) according to formulation; higher heat distortion temperatures are possible as previously noted. The highest recommended temperature for continuous use in electrical applications is usually 130°C, but suitable resins as bonds for glass or mica are used at up to 150°C. At low temperatures most resins become brittle and the limit for castings in the downward direction is set by differential contraction for the particular material and structure concerned. Epoxy–glass composites have been successfully used at liquid helium temperatures.

Water absorption is fairly low, ranging from 0·05 to 0·5%, the higher values generally applying to the filled or reinforced materials.

The resins themselves are slow burning or self-extinguishing, most of the filled and reinforced composites are self-extinguishing.

Relative permittivity of the unfilled resin generally lies between 3·5 and 5, with loss tangent in the range 0·005 to 0·03 at room temperature at all frequencies from 50–10^6 Hz, and falling with increasing frequency. Both permittivity and loss angle are higher for the flexible materials, and increase with rising temperature; loss tangent increases sharply at temperatures around 100–130°C at 50 Hz, less markedly at high frequency (Clark, 1962). High temperature curing reduces these effects. The effect of fillers and glass reinforcement is to increase relative permittivity to between five and seven, but provided the materials are dry there is no major effect on the loss tangent. Fillers may, however, worsen the effects on losses of exposure to moisture. Variations of loss tangents with temperature and with field strength in epoxy-bonded mica are given in the papers referred to in this context in Section 7.5.4.

The bisphenol resins are moderately resistant to tracking and some have even been used for periods as outdoor insulation. (The cycloaliphatic epoxies which have much less tendency to track are considered in Section 9.5.1.)

Epoxy resins are extremely useful for their high strength, good adhesion to most materials including metals (but not polythene) and resistance to moisture; the main obstacle to their wider use is their relatively high cost. Machine and transformer windings exposed to moisture or corrosive atmospheres can be coated by the fluidised bed or spraying process, or encapsulated by casting. Thermal cracking of resins cast around large pieces of metal may be avoided by using a soft, or a reinforced, resin, or by applying a flexible layer first (see Noshay, 1973). The application of cast resin to transformers and switchgear is discussed by Barker and Flack (1969). Epoxy varnishes are compatible with most wire enamels except acrylic and polyester; epoxy laminates are used for slot liners and wedges; slot insulation for small motors can be provided by simply coating the inside of the slot with resin.

In the electronic field laminates are used for printed circuit boards (IEC 249; NEMA LI–1 and DIN 40802; Harper, 1969) and potting compounds are used for components that are not too delicate. Epoxy resins have the advantage of low curing contraction imposing less strain on metallic components than other resins of comparable rigidity.

Epoxy resin systems are used for bonding mica in the insulation of high voltage machine stators, as already described in Section 7.5.4. One of the early difficulties with epoxy (as opposed to unsaturated polyester) systems was to achieve a low enough viscosity for impregnation without the use of solvents, the vapour pressure of which would cause delamination and deterioration of electrical properties during curing and service.

To evade this difficulty 'resin-rich' systems were developed in which excess of liquid resin is applied during taping, or is pre-impregnated and part cured on to the tape. For many years it was not possible to eliminate solvents entirely from such epoxy systems, and more or less elaborate processes had to be used to extract them before final curing. The earliest of these, employing an epoxy-alkyd resin, is described by Flynn *et al.* (1958). The fullest description of such a system is that given by Parriss (1971) (see Section 7.5.4) in which a formulation obtained by reacting epichlorhydrin with a novolak containing about 10 aromatic rings per molecule (see Section 9.1) is used. This is applied in a volatile solvent to mica paper backed with woven glass tape. When dried this leaves a slightly soft B-stage resin with enough flexibility to be bent without cracking; this is lapped on to the conductor assembly, pressed and cured briefly at about 160°C using a special catalyst. A further, longer, curing period follows removal from the press. Erdman and Lauroesch refer to a resin-rich epoxy tape which is completely solventless but do not give details of its composition. Present manufacturing methods require the tape carrying the B-stage resin to be flexible enough to be applied by a taping machine such as that illustrated in Fig. 47. Epoxy-novolac resin is also used for a wrapped sheet system of insulating stator bars (Denham *et al.*, 1972).

The use of B-stage epoxy resins carried on woven glass, polyamide paper or polyester paper tapes for insulating conductor bars in low voltage class F machines is growing.

Epoxy resins are available which have sufficiently low viscosity (at a temperature at which they do not cure) to be used as impregnants for mica, glass etc., tape already lapped on to stator bars; methods employed in using them are indicated by the authors quoted in Section 7.5.4.

Epoxy resins have partly displaced shellac for bonding mica V-rings for commutators.

Epoxy-mica and epoxy-glass systems are also used in the insulation of generator rotors and in dry type transformers. Tubes made from continuous glass filaments wound spirally on a mandrel with liquid or pre-impregnated resin are (Section 11.1.1) used for switchgear components, barrier bushings etc. Isogai *et al.* (1971) have described the construction of epoxy-paper condenser bushings, such as a bushing for a 275kV system having 57 foils and an effective insulation thickness of 68mm (corresponding to a mean working stress of 2.35kV r.m.s. per mm).

Epoxy-novolak resins, in which the usual bisphenol is replaced by a novolak molecule (see Section 9.1) have already been mentioned in connection with generator stator insulation. They are becoming increasingly used, and are the basis of 'single stage' resins of long shelf life and high distortion temperature (up to 280°C is claimed) used as encapsulants and moulding powders (Salensky, 1972).

Cold-curing epoxy resins have been quite widely used for filling joint and termination boxes for plastic and synthetic rubber cables, but it is being found that as the resin continues hardening over several years adhesion to the cable insulation fails and moisture can enter.

9.5.1 Cycloaliphatic epoxy resins

Another type of epoxide resin has received attention for its good mechanical and dielectric properties at elevated temperatures and for its possibilities as an outdoor insulator; it is based on cycloaliphatic, as opposed to aromatic, ring structures in the molecule. This produces resins of properties rather better than those of the bisphenol resins and less prone to tracking (see references in Section 4.5). Cycloaliphatic rings are saturated—every carbon atom has all its four valencies used—they resemble in properties the straight-chain compounds. The carbon-carbon bonds are more readily broken as in the straight-chain aliphatics, so that pyrolysis tends to produce hydrocarbon gases rather than solid carbon.

One such epoxide has the structure

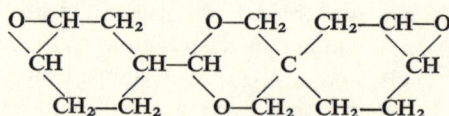

$$
\begin{array}{ccccc}
O-CH-CH_2 & & O-CH_2 & CH_2-CH-O \\
\diagdown \diagup & & \diagdown \diagup & \diagdown \diagup \\
CH & & CH-CH & C & CH \\
\diagdown & & \diagdown \diagup & \diagdown & \diagup \\
CH_2-CH_2 & & O-CH_2 & CH_2-CH_2 \\
\end{array}
$$

To realise the non-tracking properties, curing with amines or aromatic anhydrides is avoided, and hexahydrophthalic anhydride (the cycloaliphatic corresponding to the aromatic phthalic anhydride) has usually been used. Interest in this aspect of cycloaliphatics arises from the possibility of using reinforced resin insulators for overhead transmission lines. For this purpose silica and hydrated alumina fillers are being used, the latter appear to have some advantage in discouraging tracking. Field and other tests have been carried out by the CEGB and others (see Section 4.5).

Cycloaliphatic resins in the uncured liquid state have a low viscosity and greater freedom from ionic impurity than the aromatic resins and are used for encapsulation of small electronic components Salensky (1972). Some may have health hazards.

9.6 Diallyl phthalate

Whereas alkyds are polymerised entirely by esterification, and unsaturated polyesters are polymerised to linear chains by esterification then cross-linked by addition at double bonds, diallyl phthalate is a monomer consisting of an ester molecule containing two pairs of double bonds which polymerises entirely by addition:

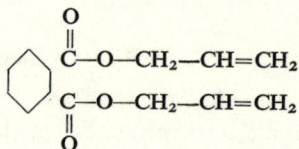

$$
\begin{array}{c}
O \\
\parallel \\
C-O-CH_2-CH=CH_2 \\
C-O-CH_2-CH=CH_2 \\
\parallel \\
O
\end{array}
$$

These double bonds are derived from the two molecules of allyl alcohol $HO-CH_2-CH=CH_2$ which condense with phthalic acid (or anhydride) to form the monomer; once this is formed no further condensation reaction takes place, and after removal of the water from the condensation reaction no other product is formed. Since the monomer is polyfunctional a cross-linked polymer is obtained.

Poly(diallyl phthalate) is a clear resin of density about $1\cdot3\,g/ml$, tensile strength $20–30\,N/mm^2$, elastic modulus $300\,N/mm^2$ and heat distortion temperature 150°C (at $18\cdot5\,kgf/cm^2$) (Roff and Scott, 1970). It is said to be suitable for continuous use up to 150°C. Permittivity is around $3\cdot5$ and loss tangent $0\cdot01$. It does not track readily.

Poly(diallyl isophthalate) is similar but with a higher heat distortion temperature and said to be suitable for use at 180°C continuously. The terephthalate is also sometimes used.

These resins are available partly polymerised as moulding resins and as solutions in solvents (such as xylene) for coating, sealing, dipping, laminating etc. The moulding material is usually filled with glass fibre or other fibre and used for small and intricate parts with small clearances, mostly in the communications field, such as radio connectors, terminal strips etc., telephone parts, switch and potentiometer casings. Mould shrinkage is said to be negligible.

The cured resin is resistant to most solvents and to weak acids and alkalis. Water absorption is low, $0\cdot1–0\cdot2\%$ for unfilled and filled resins. The glass-filled and reinforced resins are usually self-extinguishing. Glass cloth laminates have tensile strength in the region $150–200\,N/mm^2$.

A diallyl phthalate of low viscosity, suitable for low pressure transfer moulding of delicate devices, for which high strength and heat distortion temperatures are claimed, is described by Segro and Beacham (1969).

9.7 Cross-linked silicone (polysiloxane) resins

The cross-linked resins are synthetised by methods basically analogous to those for the elastomers (Section 8.14.3); a mixture of di- and tri-chlorsilane is hydrolised and condensed, the tri-functional hydroxysilane provides cross links of —Si—O—Si— like the links of the main chain. The hydrocarbon groups attached to the carbon atoms are usually a mixture of methyl —CH_3 and phenyl —C_6H_5 the latter conferring strength and heat resistance but some of the former being retained to avoid brittleness.

Varnishes for coating and impregnation are solutions of resins of molecular weight 1000–5000 in xylene or toluene, catalysts for curing, such as triethanolamine, peroxides, or ferric, cobalt or zinc naphthenate or other soaps are usually present in the solution which nevertheless has a long shelf life. In use, cure begins only after the solvents have been driven off and usually takes place at 200–250°C under a pressure, in the case of laminates, of up to $1N/mm^2$. Curing can be effected by heat and pressure without catalyst.

An after cure at 250°C may be recommended but laminates need to be constrained to prevent warping.

Moulding powder for compression moulding can be made by partially curing blocks of resin, after removal of solvent, and powdering; mineral fillers such as mica, asbestos, alumina and silica are used.

Solventless liquid resins, consisting of reactive polysiloxanes to which a catalyst is added just before use, are available for impregnation of windings.

Cured hard silicone resin has a density of about $1·6g/ml$, and a tensile strength of $20–40N/mm^2$. Silicone bonded laminates are generally weaker than polyester and epoxy but retain some strength to higher temperatures. Silicone-glass cloth laminates have densities up to $2g/ml$, tensile strength $100–200N/mm^2$, elastic modulus $1000–2000N/mm^2$, impact strength $300–500J/m$ and hardness in the range Rockwell M95–105.

The ability of silicone resin to stand high temperatures is well known; the heat distortion temperature is above 300°C and materials can be used intermittently up to this temperature. For continuous use in electrical equipment the usually accepted limit is 180–200°C. Laminates retain much of their strength after 1000 h at 250–300°C (Doyle and Harrier, 1969) and are used down to −50°C. The resin burns slowly leaving a non-conducting residue of silica. It is resistant to mineral oils, weak acids and alkalis and ozone. Moisture absorption of the laminates is 0·1–0·2%.

The relative permittivity of hard silicone resin is usually about 3·8 at room temperature and low frequency, falling to about 3·3 at 10^8 Hz while loss tangent is variable with the source of supply but is usually around 0·005 at room temperature and low frequency and does not exceed 0·03 over the measurable frequency range at temperatures up to 100°C. Above 150°C it rises rapidly to 0·1 and higher. The resins have good arc resistance and fair tracking resistance. Like the rubber they have relatively high resistance to discharge.

The laminates are used where the ability to withstand 180°C or more justifies the high cost and inferior mechanical properties: heater components, supersonic aircraft radomes and other high frequency, high temperature applications; also spacers, barriers and terminal boards in class H gas-cooled transformers (Davis and Jones, 1965) and in class H motor slot insulation. Standard properties of boards are defined in BS 3953, NEMA LI–1 and DIN 7735. See also Lubin (1969) and Parkyn (1970).

The solvented and solventless varnishes are used for impregnating windings of motors, transformers with high temperature ratings, notably in traction equipment. The silicone varnishes are compatible with polyester and polyimide wire coatings.

Where silicones are used in enclosed equipment care is needed to ensure that volatiles are driven off by heating at temperatures higher than service temperature before putting into service. In the case of enclosed rotating machines with slip rings or commutators special brushes may be necessary to avoid undue wear due to the decomposition products of traces of the silicone.

9.8 Polyurethane or isocyanate resins

This family of resins has a complex chemistry, the common factor is the use of the highly reactive isocyanate $-N=C=O$ grouping. An isocyanate is usually formed by condensation of an amine with phosgene

$$RNH_2 + COCl_2 \longrightarrow R-N=C=O + 2HCl$$

R standing for any aliphatic or aromatic group. A di-isocyanate can be made in the same way from a diamine. The isocyanate reacts readily (using a basic catalyst) with hydroxyl, acid, or amine groups:

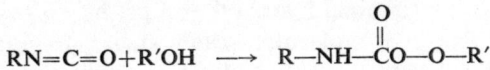

$$RN=C=O + R'OH \longrightarrow R-NH-\overset{\overset{\displaystyle O}{\|}}{C}O-O-R'$$

the urethane linkage

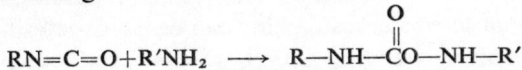

$$RN=C=O + R'NH_2 \longrightarrow R-NH-\overset{\overset{\displaystyle O}{\|}}{C}O-NH-R'$$

the urea linkage.

If a dihydric alcohol or diamine and a di-isocyanate are used the molecule can extend indefinitely. Excess of isocyanate adds a further cross-linke to the urethane linkage,

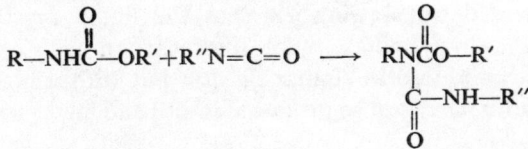

$$R-NH\overset{\overset{\displaystyle O}{\|}}{C}-OR' + R''N=C=O \longrightarrow \overset{\overset{\displaystyle O}{\|}}{\underset{\underset{\displaystyle O}{\underset{\|}{\overset{\displaystyle}{C}-NH-R''}}}{RNCO-R'}}$$

the allophanate linkage; and similarly with the urea linkage the biuret linkage is formed, and various others (see Roff and Scott, 1970; Shearing, 1972).

The group R′ may be a polymer itself, for example a linear polyester or polyether, with two or more hydroxyl groups per molecule (castor oil may be used), the group R is commonly the aromatic resulting from using tolyl di-isocyanate

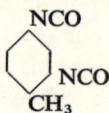

$$
\begin{array}{c}
\text{NCO} \\
\bigcirc\!\!\!\!\diagdown\,\text{NCO} \\
\text{CH}_3
\end{array}
$$

Cross-linking is obtained by using polyfunctional alcohols, amines etc., tri-isocyanates or, as already noted, excess of iso-cyanate. Toxic hazards are associated with inhaling these sub-stances and proper precautions must be taken; isocyanates of low vapour pressure and reduced hazard are available. There do not appear to be any hazards with the final polymers in any normal circumstances.

The polyurethanes have developed wide usage over the last decade (Buist and Gudgeon, 1968), principally as elastomers (including elastomeric threads from which stretch fabrics are woven), foams both resilient and rigid, and hard coatings both decorative and insulating. The latter are the most important in the electrical field.

Elastomers are produced by first linking linear polyesters, polyethers, or polyester-polyamides, of molecular weight 1000–2000 and having terminal hydroxyl groups, by means of tolyl di-isocyanate. The high molecular weight linear polymer thus pro-duced is then lightly cross-linked (vulcanised) through the iso-cyanate groups by reaction with glycols, diamines or similar difunctional substances. Foaming can be produced simultaneously with vulcanising by having some water present which reacts with isocyanate to produce CO_2

$$-RNCO + H_2O \longrightarrow -RNH_2 + CO_2$$

the resulting amine then linking to another isocyanate group as indicated earlier. The action of the CO_2 produced in this way may be augmented by blowing gas into the liquid reactants or by including a low boiling point fluorocarbon with them. Rigid foams are produced in similar fashion but with a higher density of cross links; they can be produced in situ and have good adhesion to metals.

It will be evident that polyurethane coatings can be made with a wide range of properties from flexible to rigid, for purposes such as fabric coatings and wire enamel. The usual process involves two components, one the isocyanate and the other the partly polymerised resin with suitable reactive groups. A wide range of curing conditions is available from room temperature upwards, the commonest between 100 and 150°C. Single component systems contain a 'blocked isocyanate' adduct in which the isocyanate is loosely combined, for example with phenol, and can be released by baking the coating at 150°C. Room temperature single component systems in which the isocyanate is activated by the action of atmospheric moisture are also available.

Elastomers can be in a form which can be milled and compounded with conventional rubber techniques, or in a form which can be extruded and moulded like a plastic. Sheet is made in thicknesses of 0·1–1 mm, it has a density of 1·05–1·3 g/ml, a tensile strength of about 40 N/mm^2 and remarkably good abrasion and tear resistance, with outstanding toughness. Materials used for encapsulation are cured at various temperatures from ambient to 120°C and are chosen for ease of processing or avoiding strain rather than for mechanical strength.

The upper temperature limits for continuous use range from 100 to 150°C, the lower limit of serviceability is about −30°C. Rigid foams used for encapsulation tend to soften at 100–120°C according to Clark (1962). Foams generally burn slowly when ignited.

Glass fabrics coated with polyurethanes are often recommended for use at 155°C, a more conservative figure of 130°C may be desirable in some cases. Other fabrics such as polyethylene terephthalate are also coated with elastomer.

Wire coatings of the normal polyurethane type allow soldering without previous baring of the wire; these are suitable for continuous use at 120°C and are compatible with most impregnating varnishes except silicones and polyimides. Non-solderable coatings, made by modification with polyamides, phenolics and other resins, have better heat resistance. Coatings with nylon or polyvinyl butyral 'overcoats' for heat bonding after winding are also available.

Polyurethane coatings are generally more resistant to moisture than oleo-resinous ones, and heat-cured polyurethane materials generally have low moisture absorption. The softer materials may, however, be degraded by exposure to steam or hot humid conditions. They have good resistance to chemicals and oil, petrol and ozone.

The permittivity and loss tangent are rarely of interest and vary considerably from one type to another. At room temperature and 50 Hz the loss tangent is normally about 0·01 but rises rapidly at temperatures from 75 to 120°C, depending on the type of material. Data quoted by Clark (1962) suggest that at least one elastomer has a transition temperature at about −20°C, at which relative permittivity rises from about three to over six; and some polyurethanes have a loss tangent of 0·1–0·2 at room temperature and low frequency. One would of course expect the corresponding transition for hard materials to be well above room temperature.

Resistance to tracking is good.

The major uses in the electrical field are for wire coatings and encapsulation of delicate components. The good resistance to abrasion and ability to 'solder through' are responsible for a very great interest in polyurethane wire coatings, especially for fine wires. Two-component polyurethane resin systems are used for filling junction boxes for synthetic rubber cables, the resin providing good adhesion and sealing to the rubber. The boxes are made from acrylic resin, and low vapour pressure isocyanate, which can be handled outdoors or in good ventilation without hazard, is employed. Rigid exclusion of moisture during filling is necessary to avoid foaming.

Chapter 10

Liquid insulants

Any oil, whether of synthetic, mineral, vegetable or animal origin, if free from water, acids, alcohols, amines and similar conducting liquids, will withstand the minimum electric strength tests prescribed for insulating oils. As with solids, however, the important factor is ability to maintain insulating properties and other desirable attributes in service. It has long been established that, of the naturally available liquids, only suitably refined mineral oils have the necessary chemical stability for general use. Various synthetic liquids are also sufficiently stable: chlorinated aromatics, fluorocarbons and silicones, together with synthetic hydrocarbons similar to the natural ones. Each of these has found uses in specific circumstances, but none shows signs of displacing the considerably cheaper mineral oils to any major degree.

Oil-cooled power equipment (such as transformers and oil-filled (barrier) bushings) requires a medium which is fluid at all climatic and operating temperatures, not readily inflammable, and sufficiently stable against oxidation by air not to produce corrosive substances nor to increase in viscosity nor to produce large quantities of sludge during service.

High voltage, hollow conductor, impregnated paper cables have broadly similar requirements, together with low loss angle and permittivity; service does not involve intentional exposure to air so that stability to oxidation is of less importance, but electric stresses are higher so that oils which produce gas when subjected to discharge must be avoided. Oxidation during processing must be avoided since the acids produced tend to accelerate the degradation of paper.

Solid impregnated cables require a very viscous impregnant to prevent 'draining' from high levels to low levels during service; this consists of mineral oil with thickeners (Sections 7.6.3 and 10.4).

Switch and circuit-breaker oils must maintain their fluidity under all climatic conditions and must not produce toxic or highly corrosive substances when exposed to an arc. This rules out chlorinated hydrocarbons which produce hydrogen chloride in an arc.

Capacitors may also involve high stresses (at relatively low total voltage) and require a low loss angle unless used for d.c. storage; a high permittivity is very desirable but stability to oxidation is of less importance than avoidance of gassing at high stresses.

10.1 Mineral oils

Mineral oils from different sources have different compositions, each containing large numbers of molecular species many of which are not identified. They are broadly characterised by the relative proportions of three classes of hydrocarbons: paraffinic (straight and branched chain saturated hydrocarbons), naphthenic (cycloparaffins) and aromatic (ring molecular structures with approximately equal numbers of carbon and hydrogen atoms). The proportions of unsaturated bonds and of various types of impurity are also very important.

The production of satisfactory insulating oils is partly dependent on choosing suitable crude stocks and partly on refining techniques in relation to the stocks being used. Though there have been changes in refining methods, such as the change from acid refining to solvent refining, no major change in the characteristics of the product has occurred for several decades and attention has become concentrated on the more marginal improvements and advantages, and on the techniques of ensuring various aspects of stability by quality control tests, the more important of which are summarised in the following sections.

10.1.1 Standards and tests relating to mineral oils

The abbreviations used for standard test specifications below are the same as those in Chapter 6; in addition IP stands for the Institute of Petroleum's 'Standard methods of testing petroleum and its products'. Some national and international standards prescribe not merely the tests to be used but the acceptable limits of performance in these tests; these standards include:

(i)   for transformers and switchgear, IEC 296 and 296A, BS 148, D 1040, VDE 0370, DIN 51507, (BS 148 is under revision to accord with IEC 296);

    (ii)   for cables, D 1818, D 1819;

    (iii) for capacitors, D 2297; (D 2296 refers to polybutene).

Typical values of properties are given by Pilpel and Reynolds (1960) mainly for cables and capacitors and by Clark (1962) for cables, transformers and switchgear in US practice.

Some of the methods of test used for oils have already been indicated among the general methods included in Chapter 6: flash point and pour point (Section 6.2.4), resistivity (Section 6.3.2), permittivity and loss (Section 6.3.4.2) and electric strength (Section 6.3.6.2). Tests more specific to mineral oils are indicated below, a number of new international specifications are also being prepared.

10.1.1.1   Viscosity   [BS 188, IP 71, D 445, DIN 51561, DIN 51562 and (at $-30°C$) DIN 51569]

This is controlled to ensure free flow in transformers and switchgear and appropriate impregnation and flow behaviour in cables and capacitors. (D 88 and D 2161 define Saybolt viscosity and its relation to kinematic viscosity.) Low viscosity and high flash point are, of course, opposing requirements. (The 'viscosity index' is a measure of rate of change of viscosity with temperature.)

10.1.1.2   Acidity and neutralisation value   [BS 148, IEC 296, D 664, D 974, IP 1, DIN 51558]

These are checked in new oils to ensure the absence of inorganic acids and in used oils, or oils after oxidation test, to measure the degree of deterioration. High acidity in used oils may lead to corrosion and encourages water absorption; it may indicate that sludge has also been formed, or will soon be formed.

10.1.1.3   Saponification value   [BS 148, IP 136, D 94, DIN 51559]

This measures the quantity of esters and other neutral organic substances which may result from oxidation or contamination.

10.1.1.4   Corrosive sulphur test   [BS 148, D 1275, DIN 51353]

This is to ensure the absence of sulphur or its compounds in a form which will attack copper or other metals.

10.1.1.5   Interfacial tension   [D 971, D 2285]

This is a simple measurement of the surface tension at an oil–water interface; this becomes lower when oxidation of the oil has been in progress, but the change is not quantitatively related to more direct tests for oxidation products.

10.1.1.6 Acidity and sludge of transformer and switch oils after oxidation [BS 148 (under revision), IP 56, IEC 74, D 2440, DIN 51554]

These standards require a particular procedure for heating the oil in the presence of oxygen or air and copper to be carried out, different for each of the specifications, followed by separating and weighing the sludge and titrating the acid. The various oxidation procedures represent various compromises between accelerating the effects of conditions in a transformer sufficiently to carry out the test in a reasonable time, and distorting the relative effects on different oils by too drastic a change of conditions. It is expected that the present oxidation procedure of BS 148 will be replaced by one similar to that of IEC 74.

A more elaborate group of tests has been proposed by Wilson (1965); an accelerated (48 h) test is supplemented by three longer (1000 h) tests in which the effects of oxygen access, copper catalysis and temperature respectively are severely reduced, one at a time, to enable the response of the oil under a more normal degree of influence of each of these factors to be gauged.

10.1.1.7 Acidity, resistivity and loss angle of cable and capacitor oils after oxidation [D1934]

This involves a different oxidation procedure from those of Section 10.1.1.6 followed by the acid titration and electrical measurements.

10.1.1.8 Aromatic content [D 2140]

This is estimated in a number of ways; e.g. from a combination of refractive index, density and viscosity, by liquid chromatograply, by infra-red spectrometry or by ultra-violet spectrometry. It relates to the total proportion of molecules containing aromatic groups, rather than to the proportion of aromatic groups themselves.

10.1.1.9 Content of antioxidant

This can be directly estimated by a colourimetric test [D 1473] or by infra-red absorption [D 2668] in the case of the commonest antioxidant 2, 6-ditertiarybutyl-*p*-cresol. D 2112 describes a bomb oxidation test for determining the induction period of an inhibited oil (see also Section 10.1.2).

10.1.1.10 Moisture content

This can be roughly detected at a level of 50–100 p.p.m. by the 'crackle' test [BS 148]. The oil is rapidly heated to boiling point in the quiet surroundings, any suspended water droplets cause a slight crackling sound. D 1315 describes a method of driving off water and absorbing it in weighed phosphorus pentoxide. Generally methods based on the Karl Fischer principle (Section 6.6.2) are most satisfactory and are readily adapted to oil measurement (Waddington, 1959).

10.1.2 Factors affecting the oxidation of mineral oil

It has been suggested that very pure hydrocarbons are very stable to oxidation, but if this is true the corollary that the most highly refined natural oils will have the greatest stability does not hold. In particular refinement by treatment with sulphuric acid produces a pure colourless oil which is more easily oxidised than before treatment. Stability to oxidation seems to be helped by the presence of certain natural components. Electrical properties on the other hand are generally improved by oil refining processes so long as the oil is new.

Oxidation is catalysed by copper and to some extent by other metals, it is also probable that any imperfectly cured paint or varnish immersed in the oil promotes oxidation and sludging (Norris, 1963); varnish impregnation is not now used in oil-filled transformers. Childs and Stannett (1953) found that iron and copper naphthenates in solution in oil promote oxidation at about the same rates, Massey (1952), however, concluded that metallic copper is much more effective in promoting oxidation than metallic iron or any other substance normally used in transformers, both under accelerated test conditions (BS 148) and in long period laboratory tests at 80°C exposed to air. He also found that various fully cured oil resistant varnishes did not accelerate oxidation and confirmed that protective varnish coatings on copper were an advantage.

The question whether copper catalyses oxidation at the metal–oil interface, or only after it has been dissolved by organic acids, has not been settled; very small quantities in solution, a few parts in 10^6, are sufficient to have an effect, while no additional acceleration is produced by increasing the proportion above about 1 in 10^4.

Cable and capacitor oils are generally not exposed to air except during processing; transformers may have the access to air limited by a conservator but complete isolation from air, using a bellows or a nitrogen cushion to permit expansion and contraction is expensive and not widely used.

The aromatic content of an oil increases its tendency to absorb oxygen when subjected to discharge, and was widely believed to increase its tendency to oxidation. Wilson (1965) concluded from an exhaustive laboratory study of the relative stabilities of oils under differing conditions of oxygen access, copper catalysis and temperature, that there are optimum values of aromatic content differing for each condition and for each oil, but that overall a proportion of 8–16% of aromatics (depending on the oil) is desirable. [Melchiore and Mills (1967) concluded that aromatics improve the oxidation stability if sulphur compounds are first removed.] Generally transformer oils contain 5–15% of aromatics while cable and capacitor oils which are less vulnerable to oxidation, but which should not evolve gas under the influence of discharge (Section 10.1.3), contain 15–25%.

The use of oxidation inhibitors in transformer oils has been fairly common practice in the USA for some years. The best known are 2,6-ditertiarybutyl-*p*-cresol (DBPC)

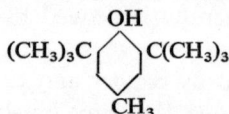

$$OH$$
$$(CH_3)_3C \bigodot C(CH_3)_3$$
$$CH_3$$

2,4-dimethyl-6-tertiary-butyl-*p*-cresol (24M6B)

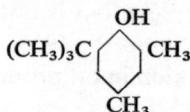

$$OH$$
$$(CH_3)_3C \bigodot CH_3$$
$$CH_3$$

and phenyl-β-naphthylamine (PBN)

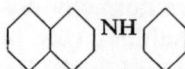

$$\bigodot\bigodot NH \bigodot$$

These additives are used in amounts of a fraction of one per cent. In laboratory tests the effect may be to reduce the rate of oxidation, but more commonly it is to produce an induction period, i.e. a period during which little or no oxidation takes place, followed by oxidation at the rate at which it would have occurred in the absence of an inhibitor; sometimes both effects are produced; the effects are characteristic of particular combinations of oil and inhibitor rather than of either independently. Massey and Wilson (1958) studied these effects by observing the absorption of oxygen at 120°C in the presence of dissolved copper using an apparatus described by Button and Davies (1953) in which oxygen is generated electrolytically as the oil uses it up. They concluded that, bearing in mind the different patterns of inhibiting behaviour already mentioned, a suitable test for inhibited oils would be to measure the time taken, under standard conditions, for the oil to absorb the equivalent of 300 ml of oxygen per 100 g of oil, which broadly corresponds to the production of an acid value of 1 mg of caustic potash per gramme of oil, this they called the 'useful life period'. This is a comparative period of the order of hours resulting from an accelerated test, not the expected life of the oil in service. They found that the useful life period is approximately proportional to the quantity of inhibitor used; that it becomes very long at temperatures below about 115°C for inhibited oils and below about 85°C for uninhibited oils; and that the addition of fresh inhibitor before the end of the useful life period prolongs it but that mixing inhibitors or inhibited oils may reduce the inhibiting effects.

While there is no doubt that inhibitors retard the oxidation of oil under the laboratory conditions most convenient for measurement, the presumption that they confer some substantial advantage in normal transformer operation does not seem to have been conclusively verified. Field tests carried out on three oils, two of them inhibited, in 17 distribution transformers operating over the period 1952–62, admittedly not representative of the best modern practice with inhibitors, gave inconclusive results (Massey and Romney, 1965). A series of simulated service tests in 10 kVa transformers by Adler and Cosgrove (1965) gave little indication of preference when operated as conservator transformers at 95–105°C although appreciable oxidation took place; only when some of them were converted to free-breathing transformers were considerable differences developed in a few months, the uninhibited oils producing the highest acid and sludge. Hornsby, Irving and Patterson (1965a) operated two 50 kVA free-breathing transformers at oil temperatures of 100–105°C, one with inhibitor, the other without. The uninhibited oil became acid after 12 months and was renewed, the inhibited oil was not; after 28 months both oils were highly acid and inspection of the inhibited transformer having had one filling showed it to be in similar condition to the uninhibited one having had two. Shroff and Wilson (1967) in laboratory tests aimed at studying oxidation with limited oxygen access, absence of copper or temperatures not exceeding those of normal transformer operation, concluded that inhibitors are of major effect when there is ample oxygen access and a large copper to oil ratio (i.e. in conditions which lead to a rather short oil life anyway).

Inhibitors are therefore most likely to be of advantage in distribution and similar transformers operating under rather severe conditions. Some inhibitors of high molecular weight are known,

H

which do not volatilise at temperatures up to 200°C, but there has been no serious attempt to use these to enable transformer oils to run at temperatures sufficiently above 100°C to justify replacing paper insulation by more expensive materials able to stand such high temperatures in oil.

10.1.3 Dissolved gases, and gases generated by discharge and by high temperatures

Insulating oils dissolve about 10% by volume of air at atmospheric pressure, this increases by a further 2% over the operating temperature range. As usual the solubility is proportional to the pressure of air above the liquid. In apparatus communicating with the atmosphere the pressure remains almost constant and the reduction of solubility with falling temperature is too slight to cause bubbles to form when the oil cools. In sealed equipment, cables, capacitors and sealed transformers, a fall in temperature results in a fall of pressure also, and gas dissolved in equilibrium at atmospheric pressure may form bubbles causing discharge and deterioration in highly stressed impregnated insulation. Oil for such equipment must therefore be degassed before impregnation. (In the case of sealed transformers with a nitrogen-filled space this is not possible, of course.)

Degassing does not eliminate the problem of spontaneous evolution of gases from hydrocarbons due to decomposition under discharge. This occurs with saturated (paraffinic and cyclo-paraffinic) hydrocarbons and has the obvious effect in highly stressed insulation of producing an unstable situation in which any discharging bubble tends to grow larger from the gases it produces itself. Unsaturated hydrocarbons on the other hand, olefines and aromatics, evolve less gas or no gas when subjected to discharge in contact with an inert gas such as nitrogen, and tend to absorb it when the same is done under hydrogen. (Under an atmosphere of air oxygen is absorbed as might be expected.) In cables and capacitors it is desirable to use blends of oil containing sufficient hydrogen-absorbing constituents to ensure that gas is not evolved, or to add synthetic ones such as polybutylene or dodecyl benzene. In transformers, oils containing a high proportion of aromatics were once considered unnecessary to prevent gassing (because electric stresses are lower), while possibly harmful in facilitating oxidation. Now it is thought that aromatic contents of 10–20% may be optimal for reducing oxidation (Section 10.1.2), and in high-voltage current transformers and bushings the higher end of this range may be preferred.

There are many types of equipment for testing gassing properties. In some the oil surface is subjected to a discharge from a metal electrode hanging above or dipping into the liquid (D 2298; Hornsby, Irving and Patterson, 1965b); in others the oil is held in an annular space between concentric glass tubes and subjected to a 'silent' discharge from electrodes on the outer surface of the outer tube and the inner surface of the inner one (see D 2300). There are many variants of each and many procedures and methods of measuring change of gas volume. Reynolds and Black (1972) outline the present situation and describe a series of experiments demonstrating the respective behaviour of saturated and unsaturated hydrocarbons. They confirm the conclusions of other authors that the rate of absorption or evolution varies (and may even reverse) with change of temperature or electric stress. They also find that ester and silicone liquids behave in a manner analogous to hydrocarbons, those containing aromatic groups tending to absorb hydrogen.

Discharges of the 'cold' type, corona, glow or partial discharge, produce predominantly hydrogen and methane from oil, but arcing under oil produces significant quantities of acetylene, while contact with metal at 200–300°C produces a variety of hydrocarbons in small amounts. Similar faults involving paper insulattion produce carbon monoxide and carbon dioxide in larger quantities than are produced from oil and its dissolved air only. Such observations led to the development of techniques for diagnosing faults in transformers while they are in operation, originally from the gases evolved and collected in transformers fitted with a Buchholz relay (Arlett, 1965). Modern methods of detecting and identifying very small quantities of gases have enabled analyses to be carried out on the gases dissolved in the oil giving more sensitive and reliable indications of faults (Dörnenburg and Gerber, 1967; Davies, 1971; Waddington and Allan, 1969). Waddington considers that the products of partial discharges at fairly low level may be detectable, and also indicates how a relatively harmless sparking between metal components may be distinguished from a fault involving destruction of insulation. Diagnosis and prognosis of faults in this way requires considerable experience of the patterns of analysis encountered in practice.

This technique can sometimes be used for locating faults which have occurred, for example in a transformer with separate tapchanger tank or in testing an oil circuit-breaker with oil-filled bushings; in such cases the failure can be located as having been in a particular tank or in one of the bushings.

10.1.4 Solution of water in hydrocarbon oil

Kaufman *et al.* (1955) found that at 25°C the equilibrium moisture content of oil is proportional to the humidity of the air in contact with it (and therefore to the partial pressure of water vapour at its surface, in accordance with Henry's law); it is usually assumed that this is true at other temperatures. They also measured the water content in equilibrium with saturated air at various temperatures from 10 to 80°C for a transformer oil and for a highly aromatic oil; these results are represented by the two lines on the graph of Fig. 48. Other measurements generally fall between these two lines, those on transformer oil being near the lower one. Increase in aromatic content, oxidation, or imperfect refining increase the water solubility, and, of course, suspended cellulose fibres will take up moisture and increase the apparent solubility, so that there is no single unique solubility curve for

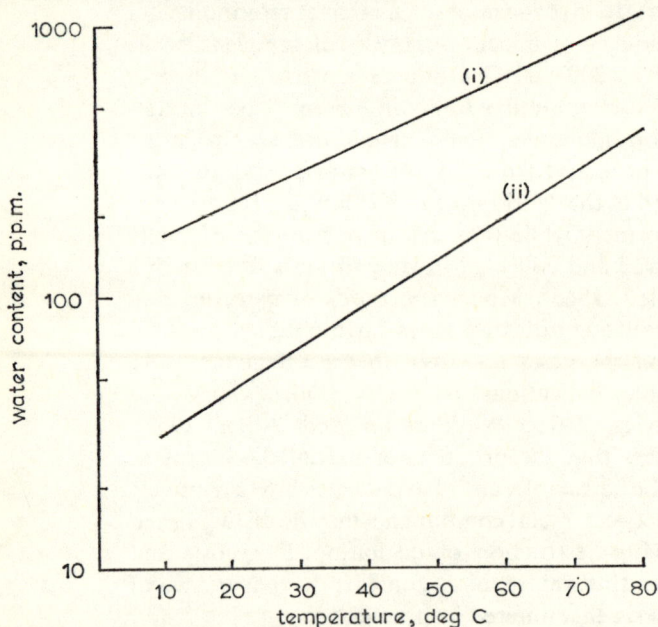

Fig. 48 Typical plots of solubility of water in insulating oil against temperature
(i) Oil with about 30% aromatic content
(ii) Paraffinic oil with low aromatic content
(Kaufman *et al.* 1955)

transformer oils. In Fig. 49 the 'oil' lines correspond to the solubility shown by line (ii) of Fig. 48 at the saturation vapour pressure of water at each temperature, combined with Henry's law to give the solubilities at lower partial pressures of water vapour. These lines will be taken as typical of a transformer oil for the discussion which follows.

The equilibrium of water content between air, oil and paper in a transformer is a topic which can be presented in a variety of ways; this, together with the differing properties of different oils and papers can lead to confusing and even erroneous explanations. In Fig. 32 was shown the range of equilibrium between moisture content of paper and humidity of air at the same temperatures. To avoid the complication of dealing with such a range Fig. 49 shows typical curves of water content against vapour pressure for a single paper, these are taken from Mason and Garton (1959) and fall within the range shown in Fig. 32. The moisture contents of oil and paper will of course be in equilibrium with each other at a given temperature when they correspond with the same vapour pressure at that temperature, whether or not there is a surface in contact with the free vapour. The kinds of processes that can take place in a transformer can be illustrated from Fig. 49.

In a transformer with paper containing 1% of moisture (conveniently, but somewhat unrealistically assumed to remain uniformly distributed in the paper), running at 60°C the water content of the oil would be 15 p.p.m. according to Fig. 49 and at this value the oil in contact with the external air would be in equilibrium with it if the air were at 20°C with a relative humidity of 60%. If the transformer were then run at 80°C the equilibrium vapour pressure of the water in the paper would go up to about $4 kN/m^2$, the water content of the oil would be brought to about 34 p.p.m. and the oil would begin to lose water to air at 20°C of any humidity. Thus the transformer would tend to dry out albeit very slowly. If now it were left idle in an atmosphere at 20°C and 60% relative humidity the stagnant oil at the air interface would be maintained at 25 p.p.m. moisture content while the oil in contact with paper holding 1% would be at about 1·7 p.p.m. (off the bottom of Fig. 49). Water would be transmitted between these regions and over a very long period the paper would tend to acquire a water content of 6·5% in equilibrium with the atmosphere we have postulated, via the oil.

If, in very unusual circumstances, the paper reached this high level of moisture and the transformer were then run successfully at 80°C again the equilibrium vapour pressure would be about

water temperature corresponding to saturated vapour pressure, deg C

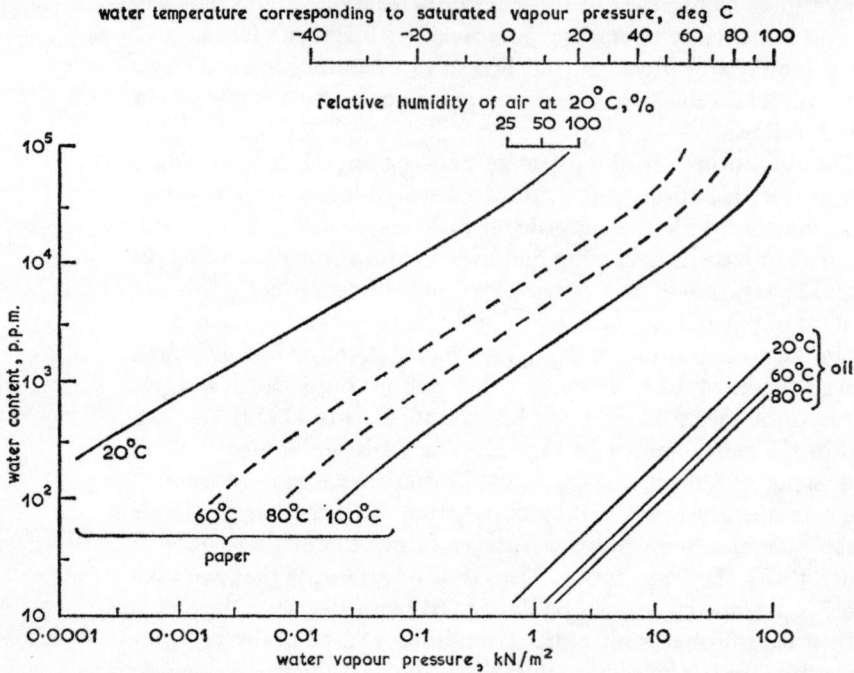

Fig. 49 Typical curves of water content against water vapour-pressure for a paper and for an oil at various temperatures

Equilibrium conditions are given by points at the same vapour pressure. The topmost abscissa scale shows the temperatures at which the saturated vapour pressure of water has the value vertically below it on the lowest scale. The relative humidities of air at 20°C having a partial pressure of water shown on the bottom scale are also shown above the graph. (Based on Kaufman *et al.* 1955; Mason and Garton, 1959)

$38 \, kN/m^2$ and the corresponding moisture in the oil would be 320 p.p.m. and would remain near this level, with very slow evaporation to the air. On switching off the transformer and allowing it to cool to 20°C water would come out of solution since at the saturation vapour pressure at 20°C (top scale on Fig. 49) only 45 p.p.m. would be in solution in equilibrium with liquid water. This water would be in the form of fine suspended droplets and if the transformer were re-energised before these had settled or been absorbed by the paper the breakdown strength of the oil might be too low to support the voltage. (Taking the transformer off load while maintaining the excitation would have a similar result.) If, on the other hand, the transformer on load at 80°C were connected to an auxiliary drying circuit (see Section 5.2.1) which held the moisture content in the oil at say 35 p.p.m. the paper would gradually be dried to about 1% moisture.

Equilibrium diagrams between water and oil can, of course, be drawn; such curves (of a sigmoid shape) are given by Binggeli *et al.* (1966).

According to Norris (1963) a power transformer in normal service operation will have a water content in the paper of around 2% and in the oil of 10–20 p.p.m. According to Fig. 49 this would correspond with a water vapour pressure of about $1\cdot3\,kN/m^2$ at a transformer temperature of about 50°C and would be in equilibrium with an external humidity of 50–100% at normal atmospheric temperatures, which is what might be expected as a mean condition.

10.1.5 Factors affecting breakdown strength

The d.c. and 50 Hz electric strength of oil under optimum laboratory conditions may be of the order of 100 kV/mm; about 50 kV/mm is usually achieved by good cable oils when tested by the standard method. Transformer and circuit-breaker oils are generally specified to have a minimum 50 Hz strength in the region of of 10 kV(r.m.s.)/mm but two or three times this should be achieved by a new oil; indeed, something like 25 kV (r.m.s.)/mm is specified by VDE 0370 for new transformer (but not circuit-breaker) oils.

According to Goswami, Angerer and Ward (1970) the breakdown strength of transformer oil remains roughly the same at frequencies from 50 to 1000 Hz.

Working stresses are, of course, much lower than test stresses, those used in impregnated paper cables, capacitors and bushings have been indicated in Section 7.1.1. In free oil, mean working stresses of the order of 1–1·5 kV (r.m.s.)/mm are common, but fields are often very non-uniform in transformers etc., and clearances tend to be decided by a study of field distribution combined with experience of what can be expected to withstand the overpotential test specified for the apparatus concerned.

It is well known that the breakdown in oil often appears to result from floating particles, especially cellulose fibres, being drawn into the region of strongest field until they form clumps or a 'bridge' of particles which, if the oil is not very dry, are likely to be moisture-bearing and slightly conductive (Ryder and Edwards, 1964; Kok, 1961; Clothier *et al.*, 1970; Watson and Higham, 1953). The use of pressboard barriers in high-voltage oil-immersed equipment helps to prevent the initiation of breakdown by floating particles, though a satisfactory quantitative account of the effects of particles or the prevention of these effects by barriers is lacking.

The combined effects of moisture and suspended particles are illustrated by Fig. 50, reproduced from Binggeli *et al.* (1966). The upper curve shows that in a filtered oil the breakdown strength is not greatly affected until the moisture content exceeds 20–30 p.p.m., which is probably the solubility limit. The oil containing 50 p.p.m. of solid particles, on the contrary, loses strength steadily as the moisture content increases from 5 p.p.m., and falls much further. Fig. 49 indicates that the water content of cellulose is several per cent for an oil content of 30 p.p.m. at room temperature, and it may be surmised that the resulting high permittivity and/or conductivity increases the tendency of particles to move into the strong field region and the tendency to cause breakdown when they get there.

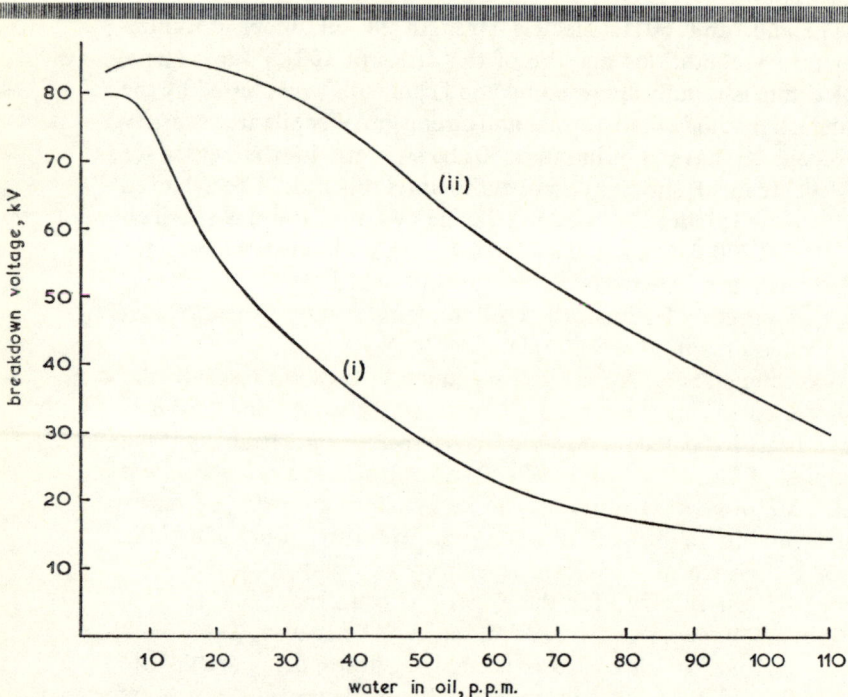

Fig. 50   Breakdown voltage of a transformer oil at 25°C between spheres 12·5 mm diameter, 3·5 mm apart

(i) Technically pure oil containing approximately 55 p.p.m. of solid impurities
(ii) The same oil after passing twice through a fritted glass filter (Binggeli *et al.* 1966)
[Curve (ii) is also similar to the curve given by Coleman (1970) for the effect of moisture on a chlorinated hydrocarbon]

Nelson *et al.* (1971) found that circulation of oil during testing raises its strength; also that the decrease of strength associated with larger gap or larger area of electrodes can be represented on a single plot against stressed volume.

Baker (1970), exploring the effect of time of stressing, concluded that the time for breakdown to occur in half of a number of specimens at constant applied voltage is inversely proportional to the amount by which the voltage exceeds a certain critical value.

10.1.6 Permittivity and dielectric loss

The relative permittivities of refined oils are in the range 2·1–2·3 at room temperature. Loss tangent at 50 Hz can be reduced to 10^{-5} at room temperature and in good cable oils should not exceed 5×10^{-5} at room temperature or 3×10^{-3} at 100°C. For transformer and circuit-breaker oils dielectric loss is not in itself of importance (increase of it is used as an indicator of deterioration, however) and may be of the order of 10^{-2} at room temperature. Loss tangent at 50 Hz and 'd.c.' conductivity usually but not always move in parallel ways, e.g. with temperature, oxidation or water content, but the 50 Hz a.c. conductivity is usually larger than the 'd.c.' (e.g. 1 min) value. This is consistent with a 'spectrum' of relaxation times spread over the range from milliseconds to minutes, much too long to be plausibly attributed to any reasonable size of dipolar molecule in a liquid of this viscosity. Conceivably an ionic mechanism of the types discussed in Section 4.2.3 and subsections may be responsible, but there is much to be learnt before such losses can be analysed. Even so simple a system as a pure hydrocarbon containing a fraction of one per cent. of an ester shows unexplained phenomena at low frequencies after exposure to air of normal humidity (Dunkley and Sillars, 1953). Mineral oil contains a whole range of substances in small proportions.

Bartnikas (1967) studied permittivity and loss in five mineral oils very fully, using wide ranges of frequencies and temperatures, and was able to draw some conclusions. In four oils of substantial aromatic content he found a relaxation mechanism operating at 10^7–10^8 Hz at room temperature moving to power frequencies at $-50°$ to $-60°$C, which could plausibly be attributed to dipolar molecule rotation and the magnitude of which was proportional to the aromatic content. (An unsymmetrical aromatic hydrocarbon molecule like an aliphatic molecule with unsymmetrically placed double bonds has a small dipole moment.) In these same four oils at 90°C he found a low frequency a.c. conductivity independent of

H*

frequency, presumably a straightforward ionic conductivity, of magnitude also proportional to the aromatic content. By addition of calcium naphthenate the conductivity was increased, and assuming (perhaps dubiously) that the salt was fully dissociated into ions, he deduced a mobility for the ions of $2 \times 10^{-9} \mathrm{m}^2/\mathrm{Vs}$ at room temperature in a liquid of viscosity 9 cP at 27°C.

A third loss mechanism appeared at temperatures in the region of 0°C and seems ascribable to a group of relaxation times of 0·1 s and greater at these temperatures, the corresponding frequencies being outside the range of measurement. Bartnikas tentatively ascribes this to 'space charge', apparently meaning interfacial surface charge, at the boundaries of a disperse phase of moisture assumed to appear only at these temperatures. (This loss is, however, much too large to be ascribed to, say, 100 p.p.m. of a disperse phase in spherical form—see references in Section 4.2.3.1, nor is there any discontinuity at the freezing point of water. An explanation might be sought along the lines of cellulose fibres becoming relatively conducting due to water transferred to them from the oil at low temperatures.)

The fifth oil, which had no aromatic content, displayed appreciable losses (0·001–0·003) at temperatures from 27°C downwards with no recognisable pattern apart from a minor probable dipole contribution. At 90°C the low frequency conductivity was nearly independent of frequency and smaller than for the aromatic oils.

The effect of increasing electric stress up to 7 kV/mm on the power frequency loss tangent of two of the oils, with and without calcium naphthenate additions, at temperatures around 25 and 85°C was also observed by Bartnikas. The increase was between 2 and 2·5 times for the most viscous oil and up to five times for the least viscous, with no very definite conclusions possible.

The effects of moisture, oxidation, contamination or imperfect refinining are always tending towards an increase of loss tangent, but it cannot be expected, nor is it found, that any of these quantities bear a single unique relationship to loss tangent except in conditions where all the others and several not mentioned remain unchanged. Any attempt to correlate electrical properties closely with, say, acid value or moisture content seems doomed to fail except in an extremely restricted context.

10.2   Chlorinated aromatic hydrocarbons

Two chemical types of substance are in general use. One is obtained by chlorinating diphenyl ⬡⬡ to the extent of 40–60 % by weight corresponding to three to six chlorine atoms per diphenyl molecule. The commercial process does not produce uniform results: for example, the 42 % chlorine material is generally called trichlordiphenyl, but it appears to be only about half made up of this compound, the other half being partly dichlor- and partly tetrachlordiphenyl. A structural formula cannot be given for any of these compounds because a mixture of isomers is produced (i.e. the chlorine atoms occupy a variety of different relative positions).

The other chemical type comprises two chlorinated benzenes, one a mixture of isomers of trichlorbenzene $C_6H_3Cl_3$ the other 1,2,3,4-tetrachlorbenzene

In the USA the term 'askarels' is recognised as meaning a non-flammable insulating liquid and seems to have become synonymous with chlorinated aromatics.

Chlorinated fluids are not suitable for switchgear because the presence of hydrogen and chlorine produces hydrogen chloride when an arc is drawn under the liquid; they are not used for cables because the high permittivity as well as the high losses would be severe disadvantages, the one adding to the reactive kVA taken by the cable, and the other to power lost. Thus they are formulated for two main purposes (Clark, 1962; Birks, 1960): transformer coolants for which the main advantage is non-flammability, and capacitor impregnants for which the primary advantage is the high permittivity. Other important characteristics are: density, which ranges from 1·4 to 1·6 g/ml; pour point; viscosity–temperature curve and volatility, all of which vary widely with the degree of chlorination. The highly chlorinated materials are viscous syrups at room temperature and become solids at well above 0°C, those with about 42 % of chlorine are comparable with transformer oil in viscosity and become solid at −15 to −20°C (rather higher than transformer oil).

Chlorinated hydrocarbons carry long-term hazards to health if ingested or if their vapours are breathed; this hazard is readily avoided in electrical applications, but its existence, together with the remarkable resistance of these and similar substances to biological breakdown processes, may lead to future restrictions on their use.

10.2.1 Transformer fluids

These are designed to have viscosities and pour points similar to, or a little lower than, those of transformer oil to BS 148. ASTM specification D 2283 defines five types, three of them blends of highly chlorinated diphenyls with chlorinated benzenes, the other two being diphenyl 42% chlorinated with and without the addition of chlorinated benzenes. IEC specifications for transformer fluids are in preparation. Because of the tendency for hydrogen chloride to be formed by unintentional discharge in these fluids they contain a 'scavenger', one of a number of materials which include tin tetraphenyl, phenoxypropene oxide and complex epoxides, in amounts from 0·1% to 0·2%.

Besides being non-flammable chlorinated diphenyls have the advantage of high thermal stability compared with transformer oils and they do not form sludge. It would indeed be possible to use sufficiently involatile fluids at considerably higher temperatures than 100°C (but they would presumably have higher pour points also) provided solid materials to stand this temperature could be used economically.

The breakdown strengths of the chlorinated hydrocarbons under normal test conditions are generally considered to be higher than those of mineral oils at power frequencies in fairly uniform fields, but lower for impulse voltages and for non-uniform fields.

Unfortunately they are not less susceptible to moisture than hydrocarbon oils, solubility of water in them is somewhat greater, and dissolved water produces a lowering of breakdown strength roughly similar to that shown in curve (ii) of Fig. 50 (Coleman, 1970).

They are powerful solvents, attacking most rubbers, plastics and varnishes. Fully cured thermohardening varnishes such as phenolic or epoxy (not modified with oil or bitumen), or cellulose ester varnishes, and polyurethane or silicone gasket materials are necessary.

In comparison with hydrocarbon transformer oils, thermal stability is a very marginal advantage, cost and weight are disadvantages to set against non-flammability (and the minor advantage of reduced transformer hum). The field of distribution in urban areas, particularly where transformers are housed within buildings, is the one where chlorinated fluids are of most use, the balance of advantages depending on local circumstances such as customs in building design, building regulations and underwriting conditions.

10.2.2 Capacitor fluids

ASTM specification D 2233 defines three types of capacitor impregnant: type A corresponds to a 42% chlorinated diphenyl with a relative permittivity of 4·7–4·9 (100°C and 1 kHz) and a pour point not higher than −14°C; type B corresponds to 54% chlorination with a permittivity of 4·15–4·35 and a pour point of 7–12°C; type C corresponds to a mixture of 75% type B with 25% trichlorbenzene—this is slightly less viscous than type A with a similar pour point and a slightly higher permittivity.

These fluids can be used for impregnating both paper and polypropylene capacitors, giving the advantages of reduced space provided the conditions are such that the higher losses are acceptable and do not involve special arrangements for the removal of heat.

All material used in contact with them must be very pure; impurities are likely to be dissolved and very small traces can produce a significant worsening of the loss tangent at 50 Hz.

In highly stressed capacitors it is important that the pour point should not be more than a few degrees above the lowest temperature the capacitor will encounter, at lower temperatures the chlorinated diphenyls become rigid and are likely to form cracks or voids which will cause damage if the capacitor is energised in this condition. The pour points are some 15–20°C above the peaks of the 50 Hz loss tangent curves (Coleman, 1964) (see Fig. 51).

Capacitor impregnants generally employ materials of 42–57% chlorination; the current tendency seems to be towards lower degrees of chlorination, in the 42% ('trichlordiphenyl') region.

Fig. 51 shows plots of permittivity and loss tangent at 50 Hz (Coleman, 1964) against temperature for various degrees of chlorination, and Fig. 52 shows similar plots against frequency at 25°C (von Hippel, 1954). These two sets are not accurately consistent because they come from different sources. Evidently, from Fig. 52, a relaxation mechanism with time constant in the region of 3×10^{-9}s for the light mobile liquid and 10^{-4}s for the most viscous one is operating, and this is consistent with dipole rotation as observed when polar molecules of comparable size are added to non-polar solutes of comparable viscosity. The temperature variation of relaxation time is also consistent with

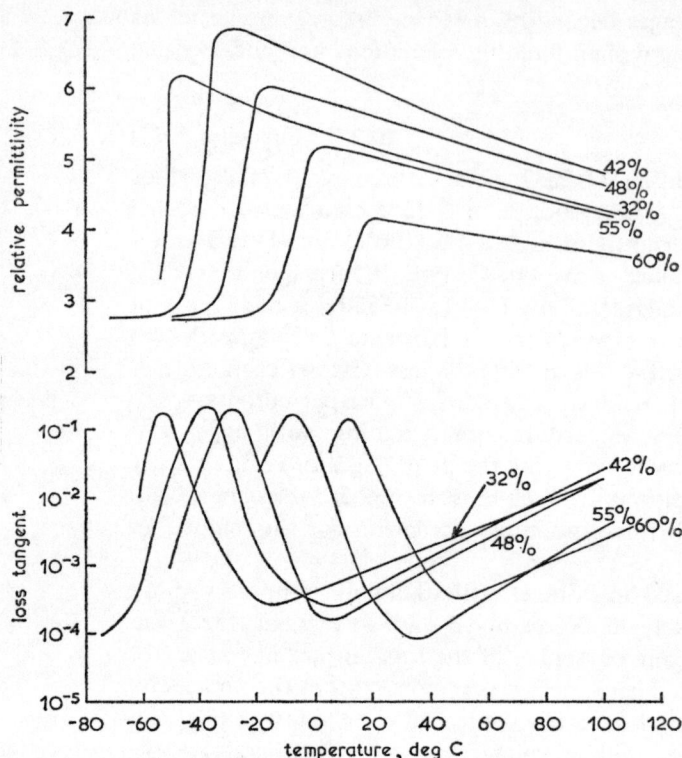

Fig. 51   Variation of relative permittivity and loss tangent of various chlorinated diphenyls against temperature

Figures on each curve indicate per cent chlorine by weight (Based on Coleman, 1964)

other known dipolar mechanisms. The permittivity 'step' in Fig. 52 is a maximum for 42% chlorination and a similar maximum appears in Fig. 51. This, as already remarked, corresponds to a mixture of di-, tri- and tetrachlordiphenyl; some such maximum is to be expected since fully chlorinated decachlordiphenyl would be non-dipolar like the unchlorinated material. The fall in permittivity above the peak temperature is to be ascribed partly to the effect of temperature in opposing the tendency of the electric field to align the dipoles, and partly to decrease in density. The high frequency permittivity is close to the square of the refractive index (1·60–1·63) indicating that there is no other dielectric polarisation operating up to optical frequencies. The dispersion curves and loss peaks of Fig. 52 are broader than those for a single relaxation time but are roughly symmetrical. Note that for any relaxation with a symmetrical distribution of characteristic times, the peak of the curve of ε'' against frequency should coincide with the point of inflection of ε' and therefore that the peak of tan $\delta =$ $\varepsilon''/\varepsilon'$ should be displaced towards the high frequencies, by an amount depending on the slope of ε'. In this material, where a large change in ε' occurs, the displacement is noticeable.

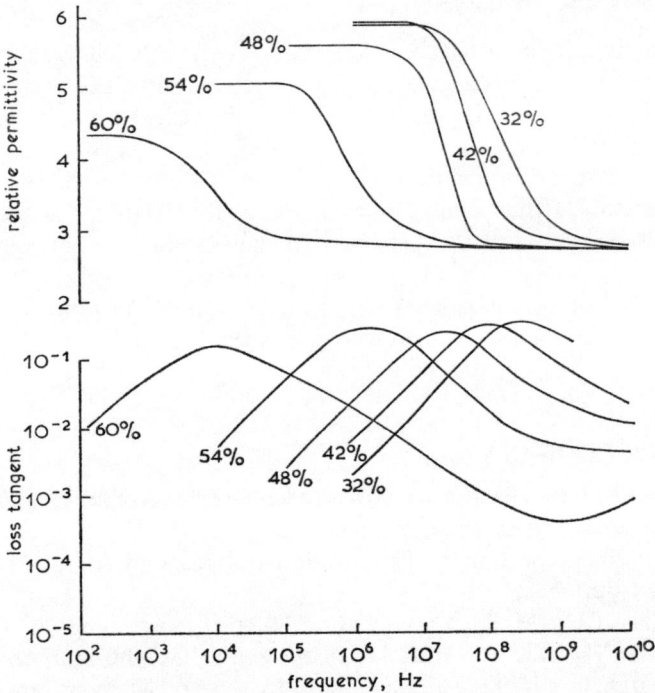

Fig. 52   Variation of relative permittivity ε' and loss tangent of various chlorinated diphenyls against frequency

Figures on each curve indicate per cent. chlorine by weight (Based on v. Hippel, 1954)

The rising loss angle at temperatures above the dipole loss peak in Fig. 51 may be ionic in origin, it appears to be a low frequency phenomenon but is not a simple conductivity.

Since Figs. 51 and 52 refer to a commercially produced mixture of molecular species none of the curves represents the behaviour of any particular substituted diphenyl; some of the individual substances in isolation have higher permittivities than any shown here; also commercial products from different sources have what different properties.

For the same reason the dielectric properties of the individual chlorbenzenes are not particularly relevant to the properties of mixtures of them with chlorinated diphenyls, these mixtures are not markedly different from those just described.

10.3   Vegetable oils and other esters

As we have seen in Section 7.7, many vegetable oils gel on exposure to air and none are so stable as mineral oils. Being esters and therefore dipolar they have higher permittivities than mineral oils and also high losses at some part of the frequency-temperature field. Castor oil finds rather limited use in impregnated paper d.c. capacitors. It has a high loss tangent at low frequencies and room temperature but it is reputed to have a high stability to discharge.

A number of synthetic esters (many of which are produced as plasticisers) are used to a small extent in a similar way.

10.4   Synthetic hydrocarbons

The best known of these are polybutylenes (or polybutenes) and alkyl benzenes (Pilpel, 1968; Clark, 1962).

The polybutylene liquids are addition polymers of the mixed isomers of butylene,
$CH_2=CH-CH_2-CH_3$, $CH_3-CH=CH-CH_3$ and
$(CH_3)_2C=CH_2$ but of much lower molecular weight than the polyisobutylene rubber which is made from the last mentioned isomer. They are available with molecular weights from about 300 and viscosity slightly higher than transformer oil to molecular weight about 1500 and viscosity 20000 cS at 37·8°C (pour point 4°C). The high molecular weight materials resemble a rather treacly rubber. They have the advantages of freedom from the naturally occurring impurities of mineral oil, but the evidence is conflicting as to whether they are more stable to oxidation (Melchiore and Mills, 1967; Pilpel, 1968). They have low losses and electrical properties similar to other pure hydrocarbons. Tables of their properties are given in the references quoted above.

They are finding use as cable and capacitor impregnants, though more expensive than mineral oils. The higher molecular weight materials are also used as thickeners for cable oil in 'solid' impregnated cable, as alternatives to rosin or wax. Precautions against gas evolution similar to those for mineral oils may be necessary.

The alkyl benzenes have properties similar to those of light mineral oils. The one most readily available is nominally dodecyl benzene (which is a 12-carbon chain with a benzene ring attached) but the commercial material only approximates to this. Its measured properties appear to be satisfactory for electrical applications, there is little experience in service.

10.5   Fluorine liquids

These are usually referred to as fluorocarbons, or perfluoro-carbons to emphasise the complete replacement of hydrogen by fluorine, but they are not all the analogues of the hydrocarbons, these are available but they are very volatile. Other substances usually included in this class are ethers or polyethers, or tertiary amines such as $(C_4F_9)_3N$. (The name amine derives from the chemical structure and does not imply the reactivity normally associated with amines.) Many of them are volatile, all are non-flammable and stable to temperatures above their boiling points. They have relative permittivities of 1·75–2·1 with low losses and good dielectric strength (Boone and Vermeer, 1970). Fluorine, like chlorine, when substituted for hydrogen seems to produce a higher a.c. strength but a lower impulse strength.

The liquids of lower boiling point, also the fluoro-chloro-ethanes, may be used for vaporisation cooling. Tables of properties of these liquids are given by Cottrell (1970) and Clark (1962).

10.6   Silicone liquids

The lower silicone polymers from molecular weight about 400 upwards are liquids. They are made in a manner similar to the rubbery polymers (Section 8.14.3) with methyl or phenyl side groups. They have densities in the region of 0·96 g/ml, viscosities ranging from a few cS upwards (at room temperature) and pour points around −50°C except for the lowest viscosity liquids which go down to −80°C (Clark, 1962). They are remarkable in having much flatter viscosity-temperature curves than other insulating liquids, low volatility in relation to viscosity, and pour points nearly independent of viscosity. The flash points of all but the lightest liquids are high (Dowling, 1960). They withstand heating at temperatures well above 120°C in air without important changes in electrical or other properties taking place (Pilpel, 1968); they do not form sludge. Out of contact with air they are stable at temperatures up to 250°C.

The silicones do not affect normal varnishes and plastics, but rubber gaskets in contact with them should be made from properly chosen materials.

Relative permittivities range from 2·2 to 2·9 [increasing with increasing molecular weight (Dowling)]; loss tangents for the methyl siloxanes are generally lower than 10^{-4} at room temperature over the frequency range 1 kHz to 10 MHz, rising to a peak of the order of 0·03 at about 10^{10}/Hz (v. Hippel, 1954; Dowling). Towards low frequencies the loss angle rises more markedly at high temperatures and at sub-zero temperatures than at room temperature but in no case does it correspond to a simple conductivity according to Bartnikas (1967). Bartnikas also examined the effect of increasing electric stress up to 8 kV/mm on the power frequency loss tangent, observing increases of three to six times.

Electric strength is comparable with that of other liquids; moisture affects it adversely.

Silicone liquids could doubtless be used for many purposes, including transformer coolants, if their cost were not prohibitive for all but very specialised applications. They are also used as water repellents.

10.7 Nitrobenzene

This liquid $C_6H_5NO_2$, of freezing point 6°C, boiling point 211°C and density 1·2 g/ml (at room temperature) is not to be regarded as a normal practical insulating fluid, but the large volume of work devoted to it over the past 10 years perhaps justifies its being mentioned here. The interest in it arises from its high relative permittivity of 35 at room temperature together with a closer approach to the properties normally required of an insulating liquid than any other readily available liquid with a comparable permittivity. If it would reliably withstand stresses comparable to those used with, say, chlorinated diphenyl, it would enable more compact capacitors to be made and probably widen their uses, and make electrostatic generation a less unatttactive proposition than at present.

Nitrobenzene as usually supplied could be described as a

semiconductor; normal purifying techniques may raise its resistivity to 10^6 or $10^7 \Omega$m, but a major improvement is produced by careful drying and purifying techniques combined with electrodyalysis, using an acid ion exchange resin at the cathode and a basic one at the anode. By this means resistivities of the order of $10^{10} \Omega$m can be attained, (see, for example, Gaspard, 1970; Felici, 1967).

The loss tangent of impure nitrobenzene at low frequencies is dominated by the conductivity, at very high frequencies there is a dipole relaxation peak.

Various experimental electrostatic generators have been constructed; Felici, Gartner and Tobazeon (1970) have made experimental capacitors, using very pure paper and aluminium foil impregnated with purified and electrodyalysed nitrobenzene. The impregnated paper had a relative permittivity of about 12, and was reported to withstand stresses of 9 and $16 \text{V}/\mu\text{m}$ at 50 Hz for 500 h. The capacitors displayed a well marked Garton effect, the loss tangent of about 0·04 at 50 Hz, 24°C (little different at 90°C) falling by a factor of seven or so as the stress was raised to $15 \text{V}/\mu\text{m}$. One capacitor which, instead of being sealed, was immersed in nitrobenzene continuously circulating through a purifying circuit, maintained a loss tangent of about 0·007 at 24°C independent of the applied stress.

10.8 Liquefied elemental gases

Apparatus working at very low temperatures, such as superconducting magnets in machines for producing high energy particles or in experimental rotating machines, must either do without liquid insulants or use the very few liquids which exist in the temperature range of interest.

Up to the present superconducting coils are effective only at temperatures of a few degrees kelvin, and are normally operated in liquid helium near its boiling point of 4·2 K. Liquid hydrogen, 14–20 K, is not cold enough for efficient use of any superconductor so far discovered, but it has been used to take advantage of the high conductivity of aluminium at 20 K and might be an adequate coolant for superconductors yet to be found. Liquid nitrogen is used in intermediate cooling stages, particularly in removing the greater part of the heat flowing down leads into the helium cryostat. Argon is included here because a substantial volume of experimental work has been done on it. Oxygen and fluorine are not considered on account of their reactivity.

Table 3 shows the basic properties of the four liquids of interest. Loss angles at low voltage are less than 1 μrad in all cases.

Table 3   Properties of low temperature liquids

	Helium	Hydrogen	Nitrogen	Argon
Boiling point, K (atm. pressure)	4·2	20·4	77·3	87·3
Melting point, K		14·0	63·2	83·9
Density, g/ml, at boiling point	0·12	0·071	0·81	1·41
Relative permittivity at boilng point	1·05	1·23	1·43	1·53

There is littled published data on working stresses and clearances in these liquids. Until recently most of the laboratory work was devoted to finding the highest stress which they would withstand when purified, distilled, filtered etc. to the greatest degree possible; this work was generally carried out with steady or quasi-steady applied voltage; the older papers are summarised by Lewis (1959) and by Sharbaugh and Watson (1962). They established that breakdown in these liquids occurs at stresses broadly comparable to those for insulating oils under similar conditions of purification but at normal temperature, and that at these high degrees of purification the effect of electrode material and condition is substantial. The work of Swan and Lewis (1960) on argon in particular established the effects of oxidation of the electrodes and of oxygen in the argon in raising the measured electric strength, effects which occur in other liquids also. Generally, electric strengths of the order of 100–200 kV/mm were found in these experiments for liquids other than helium. The measurements were mostly made with small gaps between electrodes, sometimes rejecting low values since the main interest at that time was to study the highest stress that liquids would withstand. Keenan (1972) found a wider and lower range of values for liquid nitrogen, 35–130 kV/mm, depending on gap width and electrode materials. Helium gives very erratic results, lower than the other liquefied gases; statistical studies of this are reported by Goldschvartz *et al.* (1972) and in earlier papers quoted there.

Comparative results for nitrogen, hydrogen and helium at 60 Hz under conditions more like those of normal liquid testing have veen reported by Mathes (1967) and by Jefferies and Mathes

(1970). Their results for nitrogen at 77 K and 1 atm. pressure corresponded to about 80 kV(pk)/mm, for hydrogen at 20 K and 1 atm. pressure they corresponded to 75–100 kV(pk)/mm. In both cases the electric strength rose with decrease of temperature, and could be approximately doubled by increasing the pressure to 4 atm. at the same temperature. For helium at 4·2 K and atmospheric pressure Mathes found the much lower values of 30–40 kV(pk)/mm, with electrode gaps of the order of 1 mm.

Jefferies and Mathes also measured loss tangent for nitrogen and hydrogen and found that it became measurable at about 40 kV/mm and rose rapidly to values of the order of 0·001 as the breakdown stress was approached.

Meats (1972) observed the effects of temperature and pressure on the mean electric strength of liquid and gaseous helium at temperatures above 4 K and electrode spacing of about 1 mm, finding a value of only 25 kV/mm at atmospheric pressure and 45 kV/mm at 7 atm.

This summary is only intended to indicate roughly the values of breakdown strength observed; references to the large volume of experimental and theoretical work will be found in the papers quoted.

Chapter 11

Glasses and ceramics

In this chapter the word glass is used in its everyday sense; not in the wider, derived sense employed in earlier chapters to distinguish a rubbery from a relatively rigid amorphous state. The inorganic glasses considered here, not having long-chain molecules, do not have a rubbery state.

Glasses have been regarded as good insulators from the beginnings of electrical science; ceramics were too porous and therefore too susceptible to humid atmospheres to be satisfactory insulators until, at the end of the last century, large-scale manufacture of decorative porcelains and chemical ware provided the basis for cheap impervious ceramics suitable for use as electrical porcelains.

Glasses and ceramics have a number of features in common: both are made by techniques deriving from the earliest civilised arts (though different materials are now employed); both are made directly or indirectly from minerals; both require high temperatures for their production; both are strong and rigid but fairly readily broken by impact; both can be made to have good but not, in general, superb electrical properties; both are very weather-resistant and neither tracks. Many electrical ceramics contain a proportion of a glassy phase, electrical porcelains in particular rely on this to avoid porosity. Porcelain insulators also require the smooth glass coating produced by glazing to discourage the formation of surface films of dirt and moisture.

There are several similarities in the electrical properties of glasses and porcelains. Resistivity measured by the usual d.c. methods decreases exponentially with rising temperature and is usually represented by $\rho = a \exp(-bt)$ or by $\rho = A \exp(B/T)$. The decrease for a rise of 100°C above room temperature is usually by a factor of 100 or 1000. Loss tangent similarly increases with increasing temperature in an exponential-like manner, but the increase for 100°C rise is usually only by a factor of 3–10 at power frequencies, this factor decreasing with increasing frequency down to a few per cent at centimetre wavelengths. Loss tangent as a function of frequency has very similar characteristics for all these materials, a rise at each end of the frequency scale with a featureless wide valley between them (Fig. 53)

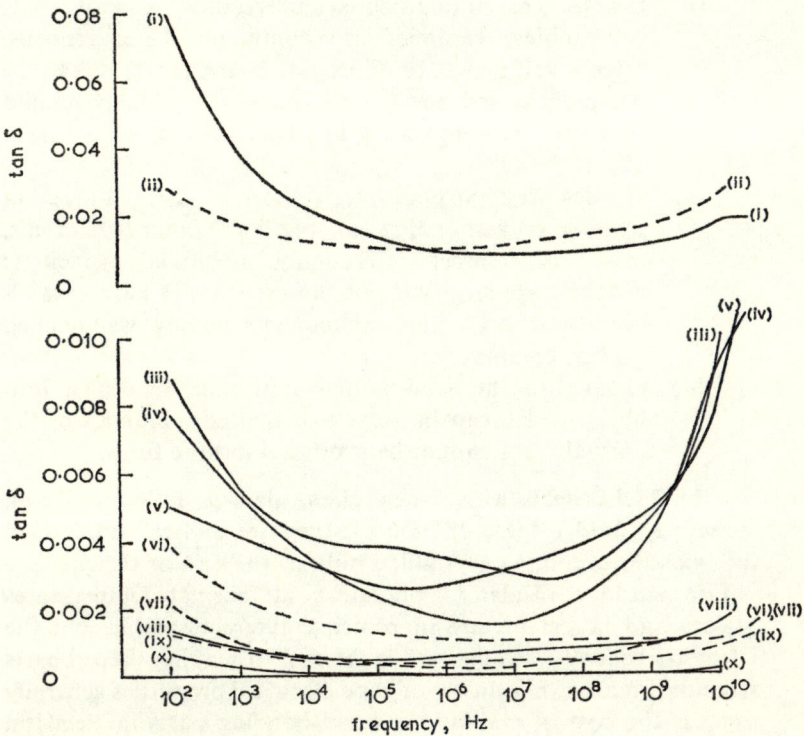

Fig. 53 Loss tangents of representative glasses (full lines) and ceramics (dashed lines), at room temperature

(i) soda-lime-silica glass
(ii) electrical porcelain
(iii) 'Pyrex' heat resisting glass
(iv) soda-potash-lead-silica glass
(v) lime-alumina-silica glass
(vi) 93% alumina, 6% silica, 1% magnesia ceramic
(vii) alumina ceramic
(viii) borosilicate glass (low alkali)
(ix) low-loss steatite ceramic
(x) fused silica

(v. Hippel 1954)

The principal differences between glasses and ceramics are:

(*a*)   Glasses must be shaped at high temperatures by blowing, pressing, rolling, drawing etc.; ceramics are shaped from powders temporarily held together by various means and subsequently heated to high temperatures so that the particles sinter together, or some components of the material melt, much of the structure remaining solid throughout and contracting but retaining its shape.

(*b*)   Glasses are homogeneous and readily freed of voids or bubbles; ceramics are commonly heterogeneous, often having two or more solid phases; they can be made dense and nearly void-free only by elaborate and expensive techniques or by the inclusion of a glassy phase.

(*c*)   Glasses are amorphous, they show no X-ray diffraction pattern corresponding to regular planes of atoms, merely a number of roughly maintained spacings; ceramics are largely (some almost wholly) microcrystalline, these crystalline components having well-defined though complex structures.

(*d*)   Glasses in the semi-molten state can be drawn into fibres, which can be woven or matted; ceramics of the normal types cannot be produced in fibre form.

Section 11.1.2 deals with a new class, glass-ceramics, which in some ways bridge these differences; they are molten and formed like glasses, becoming crystalline only at the end of the process.

For outdoor insulators operating at normal temperatures glasses and porcelains are in principle interchangeable, but the latter are generally stronger while the cost of making shaped parts from the high melting glasses of good electrical properties generally exceeds the cost of making the corresponding parts in electrical porcelain, unless they are small in size and required in large numbers.

For vacuum envelopes, on the other hand, the complications of making vacuum-tight ceramics and metal–ceramic seals are substantially worse than those of sealing leads through glass, so that glass is preferred unless the envelope must be subjected to temperatures high enough to soften available glasses.

11.1   Glasses

The individual oxides from which inorganic glasses are made have each their own crystalline forms; amorphous solids which do not crystallise on annealing and cooling can only be made from limited ranges of mixtures. To retain the amorphous structure the glass must contain a major proportion of one or more of a small number of oxides, whose normal crystal structures are such that they can be seriously distorted without losing coherence or substantially altering either the distances between neighbouring oxygen and metal ions or the energy content of the structure. Zachariasen propounded four simple rules for predicting which oxides will be *glass-forming oxides*; these criteria are now accepted (see Stanworth, 1950). The oxides which fulfil them are SiO_2, B_2O_3, P_2O_5, As_2O_5 and GeO_2, of which only the first three are of interest here. Certain other oxides, notably Al_2O_3, can take the place of one of the glass-forming oxides to some extent, reducing the amount of the latter which would otherwise be necessary. Many other oxides are compatible with glasses in moderate amounts. Among the commonest are Na_2O and K_2O, which lower the softening point and make the glass more convenient to work but often increase conduction and dielectric loss and may impair resistance to chemicals, PbO and BaO, which increase refractive index and permittivity and tend to reduce conduction and dielectric loss, CaO, MgO and ZnO which may be included for various reasons.

Practically all industrial glasses are based on silica, those based on phosphorous pentoxide and boric oxide being used for special purposes (e.g. the sodium-resistant borate glass used for sodium vapour lamps). Table 4 gives some compositions of glasses used in the electrical industry.

The structures of glasses cannot be explored in great detail, as can those of crystals, by means of X-ray diffraction patterns. These merely show diffuse bands consistent with the assumption that spacings between neighbouring ions are similar to those in the corresponding crystalline oxides but are somewhat variable so that there is no long-range regularity.

The mechanical strengths of glasses are influenced much more strongly by the form and condition of the glass than by its composition. Most glasses in sheet or rod form have breaking strengths in the region of 10–$100 \, N/mm^2$, but measured strength increases as the volume of material tested decreases. This effect is well known in most materials but is particularly noticeable and important in

Table 4   Compositions of some glasses
(per cent by weight)

Oxide	Soda or soda-lime	Borosilicate			Lead	Alkali-free alumino-silicate			Borate	96% Silica	Fused silica
							E Glass	S Glass			
SiO_2	70–74	80	74·6	65·0	57	55	52–56	65	—	96	99·5
Al_2O_3	1–2	2	1·0	2·0	1	21	12–16	25	27	0·4	—
CaO	10–13	—	0·3	—	—	13	16–25	—	—	—	—
MgO	10–13	—	0·3	—	—	—	0–6	10	10	—	—
BaO	—	—	—	—	—	3	—	—	27	3	—
B_2O_3	—	12–13	18·0	25·0	—	7	8–11	—	36	—	—
PbO	—	—	—	—	30	—	—	—	—	—	—
Na_2O / K_2O	12–16	4–5	5–9	8·0	12	—	—	—	—	—	—
Density, g/ml	2·5	2·2	2·1	2·1	2·8	—	2·5	—	—	2·2	2·2
Strain temperature, deg C	450–500	520	450	450	350–400	650	600	—	—	820	1070
Approximate relative permittivity	7 to 8	4·8 to 5·0	4 to 5	4 to 5	9	—	5·8	—	—	3·8	3·8
Uses	Bottles, windows, lamp and valve bulbs	Oven and chemical ware	Sealing to tungsten	Sealing to 'Kovar'	Lamp and valve pinches	High-pressure mercury lamps	Glass fibre	Glass fibre	Sodium vapour lamps	Low-loss, high-resistivity and high-temperature applications	

glasses because of their wide use as fibres for plastic reinforcements. The figures in Table 5 selected from results of Griffiths quoted by Stanworth (1950) indicate the magnitude of the effect.

Table 5   Tensile strength of glass fibres of various diameters

Fibre diameter, μm	108	57	47	35	29	18	14	9·6	4·2
Breaking stress, N/mm²	290	440	610	590	610	1130	1070	1600	3400

Evidence from other types of fracture indicates that this effect is primarily dependent on size, rather than on orientation as in long chain polymers. The effect of shortening the length of fibre tested is shown in Table 6 taken from results of Anderegg also quoted by Stanworth.

Table 6   Measured tensile strengths of fibres of soda-lime glass 13 μm diameter of different lengths

Length of fibre, mm	5	10	20	45	90	183	1560
Tensile strength, N/mm²	1470	1200	1190	1130	740	860	700

These effects of size of stressed specimens on the measured strength are generally explained by the presumption that strength is limited by surface imperfections. There is evidence that surface conditions such as chemical treatment, the presence of water or even the application of a film of oil have considerable effects, beneficial or otherwise, on the strengths of some glasses at least. More recently careful techniques of production and testing of fibre have tended to diminish the dependence of strength on diameter.

As with other materials the breaking strength of a glass is lower the longer the period for which the stress is applied, there may be a factor of 3:1 between fracture stress under short duration test and under continuous loading.

The theory that glass fracture is initiated at the surface is strongly supported by the results of toughening glass by cooling the surfaces rapidly from above the annealing temperature, so that the inner layers are still plastic when the outer skin has become rigid. On further cooling the outer skin is put into compression. Obviously the process must be carefully controlled to prevent such a high built-in stress that fracture becomes easier, but an improvement of three or four times in measured strength can be achieved in practice. This process is used for glass overhead-line insulators.

The long-term temperature limitations of glasses are merely those imposed by its ability to support mechanical and electrical stress; glass does not undergo progressive deterioration at high temperatures unless it devitrifies, and this does not occur with glasses in normal use. Fused silica may undergo phase changes which limit its use in certain ways. The temperatures at which various glasses may distort can be compared from the strain temperature (Table 4) which is the lowest temperature at which annealing takes place over a long period. Where a potential difference exists, particularly if it is a direct potential, the increase of conduction with temperature may set a much lower limit than does mechanical rigidity.

The electrical properties of glasses vary in complex ways with composition and also with the thermal history of the glass (Clark, 1962; Sutton, 1960; Taylor, 1960; Stanworth, 1950). Rarely does a property change monotonically with the proportion of any component, generally a maximum or minimum is reached. The same component has different effects in glasses of different types. We have already noted the generalisation that alkali metals tend to increase both conduction and dielectric loss, that the addition of some metal oxides including calcium barium, lead and magnesium may reduce the conductivity imparted by alkali metals, and that the highly polarisable lead and barium ions increase the permittivity. There are very few such simple generalisations; the more detailed behaviour is discussed at length by Morey (1954) and Taylor (1960).

Surface conduction is very sensitive to humidity (Sutton, 1960) particularly for glasses of high alkali content. Fused silica is generally reckoned to have the highest surface resistivity of all the glasses as well as the highest volume resistivity and the lowest dielectric loss.

Factors which increase the low frequency dielectric loss also increase the 'd.c. conductivity' and vice versa, and it must be strongly emphasised that, except at high temperatures, this conductivity is itself a time-varying quantity. Plots such as those of Fig. 53 could be extended downwards by several decades of frequency without reaching a region of constant conductivity where $\tan \delta$ is proportional to $(\text{frequency})^{-1}$. Much confusion of thought may be engendered by treating what is called conductivity as though it were in fact time-independent.

It is generally supposed that glasses conduct ionically and that the alkali ions are the significantly mobile ones. At temperatures where conduction is sufficiently high to measure the transport of sodium or lithium ions it is found that the amounts appearing at the cathode correspond with the charge which has passed; at lower temperatures the correlation between conduction and alkali content supports the presumption that the same ions are responsible. If direct current is passed for a long time even at high temperatures the resistance increases, as would be expected when the negative space charge near the anode, resulting from the departure of sodium ions which are not replaced, is sufficient to reduce the field strength in the rest of the space between the electrodes (Section 4.2.3.3). If an anode containing sodium is used, e.g. sodium amalgam or a molten sodium salt, this effect disappears.

It seems likely that the reason sodium or potassium ions are mobile is that they are not strongly attached to any particular fixed ion and can pass through the spaces between the oxygen ions of the silicate structure. They will encounter a series of attractions and repulsions which can be represented as a series of potential wells and barriers giving rise to relaxation polarisation of the type envisaged in Section 4.2.3.2 but doubtless with a much more complex barrier system than the double wells of Fig. 24. The tan δ curves of Fig. 53 indicate a spectrum of relaxation times growing towards longer times, and if these relaxations result from barriers of all heights it would not be surprising if the spectrum should extend to 'infinite' times. However, Taylor (1957) concluded that two soda-lime glasses, two soda-silica glasses and a 'Pyrex' glass had conductivity-frequency curves in which a d.c. component could be distinguished from a relaxation component and by subtracting the former from the latter a peak in ε'' was revealed.*
This was taken to correspond with a peak in the relaxation spectrum and occurred in the 10^2–10^4 Hz range at 400–430 K in the case of one of the soda-lime glasses. The peak relaxation time at room temperature was estimated to be 300 s for one of the soda-lime glasses and 10^4 s for 'Pyrex'.

The curves of Fig. 53 are typical of all glasses. Stevels (1957)

---

* It should perhaps be observed that even if the relaxation spectrum is rising indefinitely towards longer times this procedure will make a peak appear, since the 'd.c.' conductivity must be determined at some finite time, t_0, say. If there is in fact a relaxation spectrum extending upwards to times greater than t_0 deduction of this conductivity will discount its effects and leave a spectrum cut off at about t_0 and with a peak of at somewhat higher angular frequency than $1/t_0$. The peak should be genuine if it occurs at a much shorter time than t_0 and is independent of t_0

shows a three-dimensional model of tan δ (similar to that of Fig. 20) covering the frequency range $100-10^{10}$ Hz and temperatures from 50 to 600 K. It is bounded by rising tan δ on all sides except the low temperature one and the only peak or ridge is a small one at 50 K and about 10^6 Hz which moves to higher frequencies at higher temperatures and merges into the high frequency rise. This peak is referred to as 'deformation' loss, a term which seems to extend to any dielectric loss in the intermediate frequency range. Other terms employed are 'dipole relaxation' (for low frequency loss not attributable to d.c. conduction) and 'vibration' loss (for losses in the region 10^8-10^{10} Hz).

Since there are no peaks in the tan δ curves there is no direct evidence of relaxation times moving up the frequency scale with increase of temperature; raising the temperature merely seems to increase the loss at all frequencies, and satisfactory methods of diagnosing the causes of dielectric loss have yet to be found.

One of the principal electrical uses of glass in massive form (as distinct from fibre) is for lamp and radio valve envelopes. The choice of glass for a vacuum envelope may be governed by such factors as its suitability for automatic machine working, its softening temperature and, since glass–metal seals are required (Partridge, 1951), by its thermal expansion behaviour. Weather-resistant glasses are used for overhead line insulators and other types of outdoor insulation. The ease of obtaining very smooth surfaces with glass has led to the use of alumino-borosilicate types as bases for thin film circuits (Harper, 1969). Glasses can be used for many other components, but low impact strength has apparently prevented them from competing with ceramics or with polymers in most areas.

Glasses of various kinds are used in resistors and capacitors in 'thick film' hybrid microcircuit production (Heidler, 1969). Resistor films are commonly made from powdered noble metals (platinum, palladium, gold, silver and various alloys of these) mixed with powdered borosilicate glass and temporary organic binder to form a paste which is applied to a ceramic substrate and fired at temperatures of 750–1000°C. Capacitors of value up to a few hundred picofarads are similarly produced using glasses of various compositions and adequate atmospheric stability. A metal layer is first fired on to the ceramic base, the glass frit fired on to that and finally another metal layer added, using compositions such that the three operations require successively lower firing temperatures. The whole board with its components may be finally coated with a glaze for insulation and protection. Matching of expansion coefficients between glass and substrate is necessary in using these methods.

Since silica is here regarded as a glass (rather than as a ceramic) this seems the appropriate point at which to mention the use of SiO_2 films, grown on silicon by oxidation at 1150°C in air, oxygen or steam, to insulate the metallic control electrode (gate) from the underlying silicon semiconductor, forming a surface field-effect transistor at the interface. Similar films may be employed to isolate components such as capacitors and resistors from silicon compontents on which they are deposited. An additional layer of phosphosilicate glass has been used to prevent migration of sodium impurity from the metal into the silicon, but silicon nitride (Section 11.2.7) is more generally used.

Films of silicon monoxide, or more probably some substance intermediate between the monoxide and dioxide, produced by evaporation followed by 'annealing' in air at 400°C, are widely used in thin film capacitor dielectrics (Heidler, 1969). These are 0.5–$3\,\mu m$ thick and their dielectric properties vary according to the procedures used in making them; typically a relative permittivity of five and a loss tangent of 0.001 in the MHz range are obtained and a working stress of $60\,v/\mu m$ (about $\frac{1}{3}$ of the breakdown strength) is used. Electrodes of aluminium are evaporated before and after depositing the dielectric, any short circuit due to a defect in the latter can then be burnt out by a short impulse.

11.1.1   Glass fibre

One of the most important uses of glass in electrical insulation is in the form of drawn fibre (Waring, 1970; Mettes, 1969). For this purpose an alkali-free alumino-borosilicate glass must be used; the very large surface area of a glass cloth or mat would be too prone to the effects of moisture if it contained alkali metals. The range of compositions normally used is known as E glass, shown in Table 4, but recently a stronger fibre containing no lime, known as S glass (Table 4) has come into use. Continuous filament fibre is produced by drawing the fibres at high speed from nozzles in a platinum 'bushing' containing the melt. A number of fibres, usually 204, are grouped together in a twisted or parallel configuration to form a strand.

A small number of twisted strands may be plied to form a yarn, and a small number of yarns may in turn be plied together. Yarns are identified by the type of glass, the diameter of the individual fibres in a strand, the weight per unit length of the strand and the structure of the yarn. Strands made of untwisted fibres are grouped together and wound with little or no twist, as roving. (Staple fibre is produced in short lengths by a blowing process and spun to form yarn but is not often used in glass–resin composites.)

Continuous fibres usually have diameters between 4 and $10\,\mu m$, but smaller diameters are available; much larger diameters are sometimes used for reinforced resins of high compressive strength.

The individual fibres must be lubricated to allow them to be processed without damage, and a size or binder is also employed to hold the strands together. The lubricant may be a mineral oil, the binder dextrinised starch and gelatine or polyvinyl acetate, the total organic material may amount to 4% of the glass. Many of these materials are inimical to good electrical properties, thermal stability and the prevention of voids in glass–resin materials (Parriss, 1971). They may be removed by heating at 300–350°C and a 'coupling agent' consisting of an organo chromium compound or a complex silane can be applied to provide a bond between glass and resin (Waring, 1970). Adhesion to the glass is improved by the chromium or the Si–O grouping resulting from hydrolysis of the silane, while the 'organic' part of the coupling molecule is compatible with the resin. Careful attention to finishes of glass fibre materials is necessary to get the best electrical and mechanical properties of the composites.

Chopped strand in lengths from 3 to 50 mm may be produced from yarn and used for dough moulding or formed into mat for moulded or hand-laid composites.

Fabrics of all kinds, cloths, tapes, tubes and cords are woven from yarn and used in resin-bonded laminates, but the more interesting developments in the last ten years have been the improvements in strength obtained by using continuous filament parallel strands (roving) laid as nearly as possible along the direction of maximum tensile stress in various forms of resin–glass structure. A laminate to withstand pull in one direction only will have the strands laid in that direction, a bi-directional laminate can be made by bonding together sheets with their filaments at right angles, like plywood. Tension rods with strands running along the rod can be constructed with tapered ends of larger diameter than the centre, so that they can be anchored in conical metal end-pieces which apply the tension without local stress concentrations.

Cylindrical shapes are made by winding strands on to a mandrel (Rosato and Groves, 1964; Shibley, 1969) usually in a helical pattern; many systems of doing this have been and are being devised for different shapes and types of stress. Fig. 54 shows an early stage of one of the simplest methods of winding a tube. Here the strands are being wound dry, and will be impregnated with resin subsequently, it is preferable and usual to apply the resin to the strands, for example by running them through a bath and between rollers, as they are fed to the mandrel. Filament-wound vessels can be designed to withstand hoop and longitudinal stresses as in compressed air reservoirs, torsion stress or cantilever stress.

The resins employed for producing the various types of composite are described under the headings of the polymers concerned (Chapter 9).

Varnished glass fibre cloths have the main advantage that their thermal stability exceeds that of any of the resins used with them. The fibres do not alter in mechanical properties at temperatures up to 200°C; strength is roughly halved at 350°C. They are not suited to applications where severe flexing is involved.

Glass fibres served on to wire and treated with suitable resins are used in high-temperature windings. This type of insulation is more vulnerable to abrasion than most other wire coverings however.

11.1.2 Glass-ceramics

These materials are included under the heading of glass because the methods of making the raw material and of fabrication are those of glasses, and their compositions must include a glass-forming oxide. Any glass can, in principle, devitrify, or partially crystallise, and until recently this was merely a drawback of those compositions prone to it, which might interfere with glass-working techniques, or make old glass hazardous to drill or cut.

Supercooling of a liquid to form a glass may result from the absence of nuclei on which a crystalline phase can begin to form, or from the situation that the viscosity of the liquid at the highest temperature at which the crystalline phase can exist is so great that the rate at which atoms or ions can diffuse to the growing crystals is negligible. In the former case the glass is unstable; in the latter case a glass cooled well below the highest temperature at which crystallisation could begin will remain a glass indefinitely. In many glasses both factors operate.

Glass-ceramics (McMillan, 1964) are made from compositions that can be cooled from the melt to working temperature, pressed, blown, drawn or otherwise shaped, annealed and then cooled to room temperature without devitrifying. They are, however,

I

unstable, and contain sufficient nucleating agent to initiate crystallisation when reheated to a temperature a little above the annealing point (viscosity 10^{11}–10^{12} poise) and held there for a long period. When devitrification is complete the objects are again cooled and have the appearance of porcelain due to light-scattering from the boundaries of the microcrystals.

The nucleating agents may be noble metals such as silver, gold or platinum (or copper in the presence of a reducing oxide) which dissolve in molten glasses but tend to precipitate as colloidal suspensions when the glasses solidify. More commonly oxides such as TiO_2 and P_2O_5 in amounts of a few per cent. are used, also fluorides such as NaF and Na_3AlF_6 (cryolite). These agents are included in the melt materials, which are fused and refined by the usual methods.

The compositions employed always contain a large proportion of silica, usually 50% or more by weight, other common components are lithia, alumina, zinc oxide, lime, lead oxide and magnesia, also, of course, the nucleating agents. The microcrystals formed on devitrifying are not normally of the same composition as the parent glass, which consequently changes in composition as crystallisation proceeds; two or more crystalline phases may be formed together with a residual glassy phase.

The cross-breaking strengths of glass-ceramics are often significantly higher than those of normal ceramics, but lower than that of alumina; they fall in the range 70–300 N/mm^2. Elastic moduli are similar to those of ceramics.

Dielectric losses behave similarly to those of ceramics and glasses with change of frequency and temperature. They can be comparable with those of low loss ceramics; for example, a glass containing Li_2O, ZnO, PbO and SiO_2 quoted by McMillan (1964) has a loss tangent less than 0·0007 between 10^4 and 10^9 Hz, rising steeply at lower and higher frequencies and with a minimum of about 0·0002 at 10^7–10^8 Hz (cf Fig. 53).

One unique property of certain glass-ceramic compositions with copper or noble metal nucleating agents is that the metal can be selectively precipitated by exposure to ultra violet irradiation at the glass stage. Thus a sheet of the glass which has been irradiated through a mask, takes a latent image of the mask which can be 'developed' by the devitrification heat treatment. The practical importance is that the opaque areas thus produced dissolve much more rapidly in hydrofluoric acid than does the glass, so that holes and other shapes can be etched through the sheet for the production of printed circuit boards. Subsequently the remaining sheet can be converted to ceramic. A lithia–alumina–silica glass is used for this technique.

Capacitors in thick film hybrid microcircuits are produced from glass-ceramic compositions, the techniques being similar to those already described for glasses apart from the additional devitrifying process.

11.2 Ceramics

The general nature of ceramic manufacturing processes is probably known to the reader; their application to electrical ceramics is outlined by Popper (1960) and in more detail by Waye (1967) and by Bloor (1953). A well-documented account of electrical ceramic production is given by Eichbaum (1972).

Traditionally the shaped objects produced for tableware, decorative ware and building materials are formed using plastic clays, i.e. clays which when suitably treated with water can form a soft mass, readily shaped and joined with the fingers or simple tools, but able to retain any shape without cracking while being dried and subsequently fired. Clays are not necessarily plastic*— some of the purest and most widely used, kaolins or china clays, cannot be employed for making shaped articles without adding other more plastic clays or binders. The majority of plastic clays are impure and coloured, unsuitable for porcelains where freedom from iron and other undesirable elements is needed. Two relatively pure and highly plastic clays exist, ball-clay and bentonite, and are generally used with the china clay, quartz sand, felspar, talc or other ingredients to give them the cohesion necessary for shaping. Plastic clays are used not merely for the production of shapes from wet mixes by moulding or extrusion or the now disappearing potter's wheel; they are also used to provide the cohesion necessary for 'dry' pressing (the powdered materials containing a few per cent. of moisture being formed into simple shapes at high pressure in a steel die) and to give the strength to permit machining processes on the partly dried, shaped, body (whether originally formed wet or dry) when the required shapes are too complex for other processes.

Both kaolin and ball clay are regarded as forms of kaolinite

---

* It may be necessary, to avoid puzzling some readers, to remark that the word 'plastic' is properly an adjective describing this behaviour of clays, and the noun or adjective signifying mouldable polymeric materials as used in earlier chapters, is a derivative usage

$Al_2Si_2O_5(OH)_4$, believed to have been produced by the decomposition of felspar, typically potash felspar $K_2O.Al_2O_3.6SiO_2$, of igneous origin, by hot steam and other volcanic gases. The reason for the great plasticity of ball clay has not been simply elucidated, it is thought to be material carried by running water from the primary kaolin beds and deposited subsequently, it has a much smaller particle size than kaolin.

Bentonite $Al_5MgSi_{12}O_{30}(OH)_6$ has a still smaller particle size and its structure is readily swollen by water molecules.

Felspar itself is widely used, for example for fluxing more refractory materials (i.e. forming very local regions of a lower-melting composition which bond particles together) and as one component of the glassy phase in porcelains.

The other materials used for manufacture of electrical ceramics will be considered in dealing with particular ceramics.

Modern ceramic techniques have obvious similarities with the traditional ones, but many of the specialised materials of low loss or highly refractory character now produced for the electrical and electronic industries require the use of pure, or nearly pure, non-plastic powders such as alumina, talc, magnesium hydroxide etc. Usually the shapes required are simple plates, discs, rods and sometimes tubes. If permissible, a small amount of very plastic ball clay or bentonite is used; if not, or in addition, organic binders are used to provide temporary cohesion; these include natural gums, alginates from seaweed, poly(vinyl alcohol), starch, oils and waxes; they are volatilised or burnt off in the early stages of firing in an oxygen-bearing atmosphere.

It will be obvious that to produce a relatively dense body from the porous mixtures of clay and minerals forming the initial shapes contraction must take place during firing. This is generally of the order of 10–20% (linear), consequently it is difficult to maintain close tolerances on finished articles except in mass production of simple shapes where great uniformity of processing can be achieved.

The general properties of the ceramics in common electrical use are displayed in Table 7. Thermal expansion coefficients and thermal conductivity are included because they, together with mechanical strength, influence the ability to withstand thermal shock. Expansion coefficient also determines the feasibility of making seals to particular metals, and thermal conductivity is important in a number of applications, mostly electronic, where heat must be removed through an insulator. Cross-breaking strength is included for comparative purposes, it cannot be used to calculate

Table 7 Typical properties of ceramic materials

	Bulk density g/ml	Open* porosity %	True density g/ml	Cross-breaking strength N/mm²	Compressive strength N/mm²	Modulus of elasticity kN/mm²	Linear expansion coefficient ×10⁶	Thermal conductivity W/mK
Electrical porcelain	2·25–2·40	0·1–2·0	2·5	50–120	300–700	60–80	4–7	1·0–2·5
Zircon porcelain	3·4–3·8	0·1–2·0	3·6–3·9	140–180	600	150	4–5	4–9
Porous cordierite	2·0–2·2	up to 20	2·6	20–60	200–300	50	1–3	1·2–1·6
Vitreous cordierite	2·1–2·5	0–2	2·6	30–130	300–500	50–100	2–4	2–3
Normal steatite	2·6	<0·1	2·8	100–150	500–900	80–140	7–10	2–3
Low loss steatite	2·7–2·8	<0·1	2·9–3·1	70–120	450–600	80–140	7–10	2–3
Alumina 99%	3·9	—	4·0	400	3500	400	7·5	20
Alumina 96%	3·5	—	4·0	200	—	200	6·5	8
Beryllia	2·8–2·95	—	3·0	150–200	1600	300	7–9	50–250

* The lower figures apply to components formed from wet plastic mixes, the upper figures to those formed by dry pressing

the strength of a particular structure. Strengths of ceramics depend as much on manufacturing techniques as on composition; grain size and porosity have a substantial effect, and the strength of a particular type of structure may be affected by minor changes of shape or surface finish.

Electrical conduction in ceramics may be ionic, electronic or both. This can only be resolved at high temperatures; it is well known that zirconia is a good electronic conductor at high temperatures for instance, but mixed conduction appears to be common (Meadowcroft, 1970).

It will be noted that high permittivity ceramics based on titanates, zirconates, niobates etc. are not included in this chapter. They differ from the materials dealt with in this book in that they are used solely for capacitors and electro-mechanical transducers; in no case would one be chosen merely to separate two conducting materials and ensure that breakdown did not occur between them. They form a self-contained branch of dielectric physics; they are briefly discussed in the works by Bloor, Waye and Popper already mentioned, and more fully by Burfoot (1967), Galasso (1969), Jaffe *et al.* (1971).

11.2.1   Electrical porcelains

The essential feature of a high-voltage porcelain is that it has practically no open pores, and as few and small closed pores as possible. Porosity measured by water absorption should not be greater than 0·1 %, the total porosity deduced from the density is generally higher.

Low-voltage porcelains are not so critical (an open porosity of up to 2% is usually acceptable) and total porosity may be 5–10%.

The low porosity of porcelain is attained by using ingredients which form a glass filling the spaces between grains of crystalline material. This is called a vitreous (i.e. glassy) ceramic (not to be confused with a glass-ceramic—Section 11.1.2), but the term vitreous has also acquired the specialised meaning of a porcelain with a low porosity (defined in BS 1598), as opposed to a more porous one.

Very broadly, electrical porcelain is made from a mix composed (by weight) of one-half clay (partly ball clay, partly kaolin), one-quarter felspar and one-quarter silica, usually in the form of a quartz sand or flint. (Recently a more vigorous flux than felspar— nepheline syenite—has been introduced.) These proportions, and the use of additional constituents, vary quite widely according to

the materials available to, or preferred by, the various manufacturers. A somewhat stronger porcelain made by substituting
alumina for silica (with corresponding changes in processing) is
now coming into use for high-voltage porcelains (Ricketts, 1970).

Forming of high-voltage insulators is done by one of the wet
plastic methods; large parts are made by 'jolleying' in a rotating
plaster mould which forms the outer shape, the inner shape being
formed by tools; smaller repetition parts (e.g. suspension insulators)
may be made by quickly pressing the wet plastic body in a mould
heated to about 120°C, which gives good surface finish and clean
separation from the mould.

Firing temperatures depend on composition: English electrical
porcelains are generally fired at about 1200°C, some continental
porcelains containing less ball clay have a final firing at 1400–
1500°C. Underfiring gives too porous and weak a body; overfiring
results in distortions and various types of imperfections. Careful
control of the temperature programme is important during heating
and cooling of large components, to avoid developing thermal
stresses which may be destructive; design of the shape and section
must be considered from this aspect.

The final material consists partly of mullite crystals (Section
11.2.2.2), partly of unmelted quartz and partly of a glass resulting
from the fusion of felspar and its fluxing with quartz and mullite.

A glaze is applied to electrical porcelains to reduce the rate at
which they collect dirt, this is often applied (as aqueous suspension)
to the unfired body, its composition being adjusted so that it melts
under the conditions appropriate for firing the main body. In
continental practice the body is often fired, or partly fired, before
the glaze is applied, and then fired a second time. It is considered
that glaze improves the strength of the porcelain by filling surface
imperfections or by being in compression (due to contracting less
than the main body) so that surface cracks are less likely to be
initiated.

On surfaces where the insulator is to be bonded with cement to
form a cushion between metal and porcelain components, a special
combination of glaze with granulated fired clay is applied to give
a good grip. Portland cement is commonly used, it must be carefully prepared and after application aged in steam to prevent
'growth' over several years and bursting of the porcelain.

For many years studies have been made to develop a semi-
conducting rather than an insulating glaze for suspension insulators.
This should provide a more uniform surface stress, overcoming
the concentrations arising from the shape of disc insulators, from

pollution and from water films during rainfall, improving flashover voltage and decreasing discharge and radio interference. The glaze must, of course, have a high enough resistance not to become overheated, but mild warming may be beneficial in wet conditions. After experiments with many types of conducting oxides a glaze based on tin oxide with a proportion of antimony oxide has been developed (Ricketts, 1970) which withstands weather conditions. By thickening the glaze near the cap and the pin and by spraying a metal layer where they make contact difficulties due to concentrations of current and bad contacts are said to have been overcome.

Normal electrical porcelains are not recommended for use at temperatures above 100°C. Electric strength diminishes above this temperature, and resistance to thermal shock is poor.

The electric strength of electrical porcelains is comparable with that of good organic insulation, the slightly porous material used for low voltages has lower electric strength. BS 1598 specifies test limits corresponding to about 10 kV/mm for wet plastic formed porcelain and 2 kV/mm for pressed powder bodies. The values obtained from test specimens are not, however, necessarily consistent or useful; breakdown tests if needed should be performed on samples of components (BS 137 and BS 3297). Normally external flashover is the limiting factor, not breakdown of the porcelain.

Many different figures are quoted for the dielectric properties, not surprisingly in view of the varied materials and methods of manufacture. Broadly the permittivities lie between five and seven (the higher value applying to wet-processed and denser bodies), and curve (ii) of Fig. 53 is typical of the loss tangents of electrical porcelains at room temperature (though values twice these would not be unusual) rising by a factor up to 10 at 100°C. Resistivity is typically $10^{12}\,\Omega$m at room temperature.

The uses of high-voltage porcelains are almost too well known to mention; they include overhead line suspension and pin type insulators, outdoor weathersheds for bushings and switchgear insulators including reservoirs for gas-blast breakers. Structures up to 3 m high can be manufactured. Low-voltage porcelains are used for a great many small components but are being superseded by plastics except for outdoor applications.

11.2.2 Refractory insulators

Insulators for use in domestic cookers, radiators, resistors, furnaces etc. can usually be porous without disadvantage, and for temperatures less than 1000°C many varied ceramics are made from such materials as 'grog' (products of fired clay bodies coarsely ground), and coarse alumina, bonded with clay and fired to a high temperature. These formulations are deliberately made porous so that high temperature firing does not cause blistering and because porous materials are more resistant to thermal shock.

11.2.2.1   Cordierite ceramics

Cordierite is a substance of definite structure and composition, $2MgO.2Al_2O_3.5SiO_2$, and very low coefficient of thermal expansion (about $10^{-6}/°C$). Cordierite ceramics are usually produced from a mixture of clay and talc (Section 11.2.5) with various proportions of grog, alumina or sillimanite; they contain a proportion of cordierite which depends on the mix employed and on rather critical firing conditions at temperatures of 1400–1450°C, the Al_2O_3—SiO_2—MgO system being a complex one (see Bloor, 1953). Expansion coefficients are usually larger than that of cordierite itself (Table 7). A vitreous cordierite ceramic of lower firing temperature can be made with small additions of felspar or alkaline earths. It seems that ceramics having compositions approximating to that of cordierite, even though they contain little material of the cordierite structure, generally have low expansion coefficients and some degree of endurance to thermal shock. They do not have particularly low dielectric loss; relative permittivity is usually between four and five. They are normally used for electric heating applications at temperatures from 1000–1300°C and for fuse bodies.

11.2.2.2   Mullite ceramics

Mullite, $3Al_2O_3.2SiO_2$, and silica are the normal products of heating kaolin or ball clay to 1200°C or more. Mullite is a highly refractory material, and ceramics made by firing sillimanite, $Al_2O_3.SiO_2$, and clay were used at one time for spark plugs. They are now used for thermocouple sheaths, and their texture makes them very suitable as base materials for pyrolitically deposited carbon resistors.

11.2.3   Zircon porcelains

Zircon is a highly refractory mineral, $ZrSiO_4$, found as sand in Australia, India, Brazil and elsewhere. Zircon porcelain is produced from a mix of ball clay, alkaline earth silicate and zircon fired at 1300–1400°C. It contains a glassy phase in which unchanged zircon is embedded, and has a much higher mechanical strength, thermal conductivity and endurance to thermal shock than electrical porcelain. It is, however, relatively expensive to produce on account of the wear on dies caused by the abrasive zircon and of the high and critical firing temperature. There are a variety of formulations with relative permittivities from seven to twelve, loss tangents of 0.005–0.001 varying little with frequency and a high resistivity which remains above $10^4 \Omega$m at temperatures up to 600–700°C. They are now used for high-frequency high-strength insulators and for some high temperature applications.

11.2.4   Pyrophyllite

This naturally occurring mineral, $Al_2Si_4O_{10}(OH)_2$, is best known for its softness and machineability in the natural state and the relatively small shrinkage when fired to a hard ceramic, which requires temperatures in the region of 1200–1300°C. The ease with which it can be shaped results from the crystal structure which consists of pairs of layers of silicon and oxygen with a layer of hydroxyl ions between each pair. The electrical properties of the fired ceramic are somewhat variable, it finds uses where terre is need for a greater stability to high temperatures or a hgeater rigidity than organic materials can provide, but the number of parts required is too small to justify the production of a normal ceramic.

11.2.5   Steatite ceramics

Natural talc, soapstone or steatite, $Mg_3Si_4O_{10}(OH)_2$, has long been known and used in the same way as pyrophyllite (Section 11.2.4), the structure being similar but with three magnesium ions substituted for two of aluminium. The fired product has low dielectric loss but being porous it only retains this property when kept very dry. It is used for internal components of evacuated devices, the porous structure permitting rapid degassing.

The demand for low loss ceramics starting in 1930 led to the commercial development of dense synthetic magnesium silicate bodies, basically $MgSiO_3$, not susceptible to humidity, and the word 'steatite' now refers almost universally to these materials. They are made from mixes consisting essentially of talc with 5–15% of ball clay. The steatite described as normal in Table 7 contains a few per cent. of felspar which helps to impart a high mechanical strength but also increases dielectric loss and conduction, especially at high temperatures. Low loss steatite generally contains the minimum possible sodium and potassium but the formulation contains a few per cent of barium and calcium carbonates which form complex glasses with the silica and alumina of the clay, these acting as a low-loss bond for the $MgSiO_3$.

The normal (felspar-containing) steatites are fired at temperatures from 1300–1400°C, the low-loss steatites have a firing range 50–75°C lower. Firing contraction is 10–20%. Special precautions are necessary during manufacture to avoid instability of properties due to the presence of a metastable form of $MgSiO_3$. Endurance to thermal shock is not exceptionally good but is adequate for most purposes.

Relative permittivities of different steatite compositions range from 5·5 to 6·5 at room temperature; Fig. 53, curve (ix), shows the loss tangent of a typical low loss steatite; similar data for a selection of steatite materials are given by v. Hippel (1954). Examples of the effect of temperature are also given by v. Hippel, typically temperature rise produces a noticeably rapid increase at 100 Hz and very little increase at 10^{10} Hz. The felspar-bearing materials are likely to have a more rapid increase of loss with temperature at low frequencies than the low-loss ones. Resistivity remains above $10^4 \Omega$m up to about 400°C for the former and up to 800°C for the latter.

Steatite ceramics are widely used for small insulators of all kinds in high frequency equipment, for printed circuit and hybrid integrated circuit bases (Harper, 1969) and for trimmer capacitors. They lend themselves well to quantity production by dry-pressing methods because talc reduces die wear and because techniques for obtaining dense fired bodies from dry-pressed materials have been found. They are fairly cheap and are often used for general purposes where low loss is not important.

Forsterite, Mg_2SiO_4, closely related to steatite, is the basis of ceramics produced from mixes similar to those for steatite but with the addition of magnesium hydroxide. Forsterite ceramics have generally a slightly lower loss and higher expansion coefficient than steatite ceramics, they are more vulnerable to thermal shock but are suitable for sealing to titanium metal and nickel-iron alloys.

Wollastonite, $CaSiO_3$, is another low loss material with loss tangent in the region of 10^{-4} which can be incorporated in a ceramic bonded with a low loss glass (Bunag and Koenig, 1962).

11.2.6 Single oxide ceramics

Two metal oxides have such desirable properties as to make commercially worth while the difficult processes of producing impervious refractory ceramic bodies with little clay or flux addition. These are alumina which is produced on a large scale and beryllia which is produced for special purposes.

The possible small-scale techniques for preparing impervious bodies from pure refractory oxides have been reviewed by Archbald and Smith (1953), one of the major problems being to retain the shape of the article while it is being brought to the sintering temperature, without introducing contaminating substances. The techniques include frequent tamping in a die, the use of very high pressures, pressing in a rough vacuum, use of temporary volatile binders, 'hydrostatic' pressing (see below), pressing at high temperatures and partial firing while mechanically supporting the shape. The production of dense bodies without fluxing requires sintering temperatures within a few hundred degrees of the melting point of the oxide, the purer the material the higher the temperature.

11.2.6.1 Alumina

Pure Al_2O_3 has a melting point of 2050°C and can be sintered at about 1800 C. Dense ceramics of 99 % or higher purity are available; 95 % alumina is produced in large quantities, it is made from a mix containing a few per cent of bentonite and about 1 % of MgO which restricts the crystal growth, the sintering temperature being about 1600°C.

Pure alumina has a permittivity of 9·4 and very low loss (Fig. 53).

Spark plugs are produced by hydrostatically pressing the approximate shape (in a rubber or plastic bag surrounded by fluid at high pressure) which is then ground or turned to the final shape and fired. Some details of a typical process are given by Waye (1967).

Other uses of alumina include vacuum envelopes of high-power radio valves, microwave windows in u.h.f. generating valves, nose radomes for missiles (required to stand high temperature and ablation as well as to be transparent to microwave radiation) and thermocouple sheaths. An important new use of alumina sheet is as a base material for high-frequency printed circuits, microstrip transmission lines and hybrid integrated circuits (Harper, 1969).

The ceramic sheet is not smooth enough to use as a base for 'thin film' components, for this purpose it is glazed with aluminoborosilicate glass. If a ceramic surface with a roughness less than 0·1 μm could be developed it would be possible to dispense with the glaze. Efforts to achieve the necessary surface finish without glaze are being made (Hill *et al.*, 1970).

Alumina can be produced in single crystals (boules or synthetic sapphires) by flame fusion of a powder (Verneuil process) or by crystal pulling from a melt (Czochralski process), these have very smooth surfaces and are used for thin film circuits if the cost can be justified.

A possible alumina insulation which has been explored from time to time is the anodic film, formed on aluminium wire for example and dried. This has not so far proved nearly reliable enough for general use. The subject of anodic film capacitors will not be pursued here since it does not involve a separable insulating material.

Studies in crystal growth from the gas phase have lead to the discovery that certain materials will form single crystals, generally of diameter $1–10\,\mu m$, known as 'whiskers'. Alumina is one of these and techniques for large-scale growing are being developed, though the cost is prohibitive for ordinary purposes.

Alumina single crystal whiskers $1–10\,\mu m$ diameter can be grown from a gas phase reaction, e.g. between aluminium trichloride vapour, hydrogen and carbon dioxide at a temperature of 1700–1800°C. The tensile strength of a whisker is very high, of the order of $10^4\,N/mm^2$. It is an ideal material for reinforced resin materials and is in small scale production, but its use can rarely be justified.

11.2.6.2 Beryllia

BeO, melting point 2570°C, is regularly but not very widely used; its cost is about two orders of magnitude greater than that of alumina and its airborne dust carries hazards exceeding those of silica and asbestos; very stringent dust removal precautions are necessary at all stages of powder handling. The pure dense material has a relative permittivity of about 6·4, and low loss, but its most remarkable property is a thermal conductivity (Table 7) greatly exceeding that of any other non-metallic material and comparable with that of aluminium metal. It is used as a base material for integrated circuits and solid state devices, and whenever it is necessary to remove large amounts of heat through an electrical insulator with the minimum temperature rise.

11.2.6.3 Magnesia

MgO, melting point 2800°C, is used for ceramic bricks for high-temperature furnace linings but is not often used in ceramic form for electrical purposes, perhaps because, even when calcined, it tends to absorb atmospheric moisture. In certain circumstances it might be useful for its thermal conductivity. It is widely used in powder form in tubular heaters for domestic cookers and industrial heating and in mineral insulated cable.

In making tubular heaters a fairly coarse powder (which has been fused and reground) is packed into the metal sheath around a spiral heater filament. The tube is then flattened and swaged to a smaller cross section to compact the oxide. (An older process in which magnesium metal was converted to the oxide *in situ* is now very rarely used.)

Magnesia insulated cable ('mineral insulated cable') is made by pressing annuli of magnesium oxide with a small quantity of binder or flux and partially sintering at 800–900°C. These annuli are threaded on to a copper rod and surrounded by a copper tube or sheath, the whole assembly being of much greater diameter than the finished concentric cable is to be. It is then reduced in diameter by successive drawing operations. This type of cable requires special bending and end-sealing methods during installation but it can be used at relatively high ambient temperatures, e.g. in steel works, and the current rating for a given conductor section is about twice that for a p.v.c. insulated cable. The margin of breakdown strength for impulse voltages seems to be lower than for plastic insulated cables and surge protection may be necessary on certain circuits.

11.2.7 Nitrides

Boron nitride, BN, density 2·1–2·3 g/ml (sublimes at 3000°C) and *silicon nitride* Si_3N_4 density 3·0–3·2 (sublimes at 1900°C) have specialised uses as ceramics (Popper, 1959).

Boron nitride is normally produced in crystals similar to those of graphite (it is also called 'white graphite'). The nitride powder is produced by reacting boric oxide or boron trichloride with ammonia gas at about 900°C; it cannot be sintered normally but fairly dense bodies can be produced by the rather expensive technique of hot pressing at 1900°C. This material is easily machined, has a low friction coefficient and has a cross-breaking strength of 50–100 N/mm². It retains considerable strength to 1000°C but oxidises fairly rapidly in air above 700°C. The relative permittivity is in the range 4·1–4·5, loss tangent is of the order of 0·001–0·0001 at frequencies from 100 to 10^{10} Hz and temperatures up to 200°C. It is somewhat porous and susceptible to moisture but finds uses in specialised high-temperature applications such as missile radomes where a long life in atmospheric conditions is not required. Boron nitride films have been formed chemically.

Silicon nitride, Si_3N_4, can likewise be prepared in massive form by hot pressing the powder obtained by reacting silicon tetra-chloride with ammonia. It can be made with densities ranging from 2·8 to 3·4 g/ml, cross-breaking strengths of up to 1000 N/mm^2 and a thermal conductivity of about 8 W/mK (Waye, 1967). Mechanical strength diminishes slowly above about 800°C and is halved at about 1300°C. A more porous body obtained by more conventional sintering techniques has a reported strength of about 300 N/mm^2. Silicon nitride can be deposited as a film on a hot molybdenum or silicon substrate from a mixture of silane and ammonia (Doo *et al.*, 1968), or by sputtering (McMillan and Misra, 1970). The properties of the film depend on deposition conditions; typically it has a permittivity of eight (figures from five to twelve are reported) and a very low loss tangent at frequencies from 10 to 10^{10} Hz, retaining good dielectric properties at temperatures up to 600°C. It has been used as a thin film capacitor dielectric, and as insulant for surface field-effect transistors, deposited on top of a film of silica produced by oxidation of the semiconductor. Used in this way it prevents diffusion of foreign impurity ions from the control electrode to the silicon surface; also the combination has specific properties which may be exploited (Szedon and Stelmak, 1970).

Whiskers of silicon nitride can also be grown.

11.2.8 Metallising and sealing to metals

Methods of coating ceramics with metal film are required for the production of capacitors, for providing electrostatic screens on ceramic envelopes, for producing high frequency inductors and printed circuits on low loss ceramics, and for the attachment of a metal by soldering, brazing etc. to form an hermetic or vacuum-tight seal.

Metal flame spraying of brass, copper or zinc can be used occasionally where the ceramic has a rough surface: the coating is porous but can be made continuous by tinning with solder. Evaporation or sputtering of metals is used for capacitor electrodes, this avoids the use of organic substances but adhesion depends on surface preparation, and the production of highly conducting coatings by these means is not always economic.

Silver is the coating most commonly used except for metal-ceramic seals meant to stand a high temperature. For some purposes a suspension of silver flakes in varnish, which is cold cured or stoved at 100–200°C, is adequate. Where strong adhesion and absence of dielectric loss are required a 'fired-on' coating is used, in which silver or platinum or their salts, together with a flux such as a borate or a fluoride (to form a vitreous bond to the ceramic), are carried in an organic vehicle and painted or sprayed on the ceramic, then fired at 700 or 800°C. During firing the vehicle is volatilised or burnt and the coating fluxed to the ceramic. Many proprietary coatings are of this type.

Silver films can be soldered to leads or to metal casings (e.g. the bushings of a high voltage capacitor can be soldered to the case); a lead–tin solder containing silver is used to prevent the silver film being dissolved away.

Solders containing indium can be bonded directly to ceramics with sufficient strength for some purposes, and will also alloy with copper, brass, steel etc.

Where a ceramic is used for a high temperature vacuum device which may operate at several hundred deg.C the associated metal parts will be made from a refractory metal and the preliminary metallising must also use a metal of high melting point which will bond to ceramic and alloy with brazing metal or directly with the metal component to be sealed. Typical methods (Kohl, 1964) are to coat the high temperature ceramic (the area to be sealed will normally have been ground to produce an accurately flat or cylindrical surface) with titanium or zirconium hydride and heat in a vacuum furnace to about 900°C together with braze metal and the metal to be bonded to it. The hydride decomposes and the titanium or zirconium react with the ceramic and also with the braze metal; all this can be done in one operation. Alternatively molybdenum or tantalum powder with a small proportion of manganese suspended in nitrocellulose is applied to the area to be sealed and the ceramic heated to as high a temperature as it will stand, in wet hydrogen. The coating thus obtained is electroplated with copper or nickel and the metal component brazed to it in vacuum or a reducing atmosphere. Analogous methods employing oxides or salts of refractory metals are also in use (Binns, 1970).

The choice of metal and ceramic to be bonded is generally made on one of two principles; either the expansion coefficient of the metal and the ceramic are closely matched and the metal is soft enough to adjust to any mismatch by plastic deformation (thermal expansion is rarely linear over a wide temperature range) or the geometry is such that, on cooling, the metal contracts on to the ceramic and the joint remains in compression at all working temperatures; in this case a soft metal, such as copper, is often used.

Appendix

The transformation from eqn. 2.2 for steady applied stress to eqns. 2.3–2.6 for a sinusoidal one, and similarly from eqn. 4.3 to eqns. 4.4–4.6 can be verified by analogy with the equations for the circuit of Fig. 55. Here voltage V and charge Q will represent the same quantities in eqns. 4.2–4.6 and will represent stress q and strain ϕ, respectively, in eqns. 2.1–2.5.

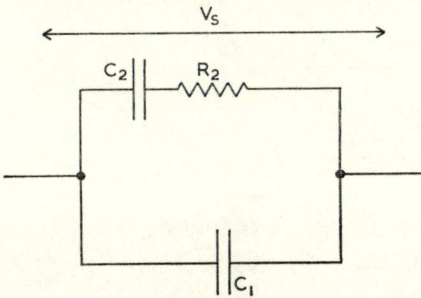

Fig. 55 Resistance capacitance analogue of a mechanical or dielectric relaxation

When a steady voltage V_s is first applied to this circuit the initial charge Q_i is the charge on C_1 which remains unchanged and the total final charge on C_1 and C_2 is

$$Q_s = Q_i + C_2 V_s \quad . \quad . \quad . \quad . \quad . \quad . \quad \text{A1}$$

The growth of charge on C_2 is governed by

$$R_2 \frac{dQ_2}{dt} + \frac{Q_2}{C_2} = V_s = \frac{Q_s - Q_i}{C_2} \quad . \quad . \quad . \quad . \quad \text{A2}$$

If we write Q for the total charge on C_1 and C_2 at any moment,

$$Q_2 = Q - Q_i$$

K

and putting $\tau = C_2 R_2$ the differential equation eqn. A2 can be written

$$\frac{dQ}{dt} + \frac{Q - Q_i}{\tau} = \frac{Q_s - Q_i}{\tau}$$

or

$$\frac{dQ}{dt} = \frac{Q_s - Q}{\tau}$$

having the solution

$$Q_s - Q = (Q_s - Q_i)\exp(-t/\tau)$$

which corresponds to eqns. 2.1 and 4.2a.

The a.c. admittance of this circuit is

$$j\omega C_1 + \omega j C_2 \frac{1 - j\omega\tau}{1 + j\omega^2\tau^2}$$

so that it can be said to have a complex capacitance $C = Q/V$ (corresponding to complex compliance J in eqn. 2.3 and complex permittivity ε in eqn. 4.4) of

$$C = C' - jC''$$

where

$$C' = C_1 + \frac{C_2}{1 + \omega^2\tau^2}$$

$$C'' = \frac{C_2\omega\tau}{(1 + \omega^2\tau^2)}$$

Using the 'initial' and 'final' capacitances C_i and C_s, $C_i = Q_i/V_s = C_1$ and $C_s = Q_s/V_s = C_1 + C_2$

$$C' = C_i + \frac{C_s - C_i}{1 + \omega^2\tau^2}$$

$$C'' = \frac{(C_s - C_i)\omega\tau}{1 + \omega^2\tau^2}$$

the last two equations corresponding with eqns. 2.4, 4.5, 2.5 and 4.6.

Bibliography and index
of authors

Numbers in italics indicate the pages of this book on which the authors are mentioned. Conference titles which have been abbreviated are marked with an asterisk and given in full at the end of the list.

ADAMCZEWSKI, I. (1969): 'Ionisation, conductivity and breakdown in dielectric liquids' (Taylor & Francis); *52*

ADLER, H. A., and COSGROVE, M. F. (1965); *IEEE Trans.*, **PAS 84**, pp. 657–66; *211*

ALLAN, R. N., and KUFFEL, E. (1968): *Proc. IEE*, **115**, pp. 432–40; *119, 145, 160, 163*

ALMOULI, J. A., and BRUINS, P. F. (1968): 'Heat resistant polyamides and polyimides for electrical insulation' *in* 'Plastics for electrical insulation' (Bruins, P. F., ed.) (Wiley-Interscience); *171*

ANDERSON, A. R. (1965): *ASEA J.*, **38**, pp. 87–93; *131*

ANGERER, L.: see GOSWAMI

APPLEBY, J. T., DAVIES, C. L., DAY, A. G., DOUGLAS, G. C. G., HOWARD, C. P., and PARRISS, W. H. (1970): *Proc. IEE*, **117**, pp. 1777–81; *30, 97*

APPS, L. T. (1970): *Electronics & Power*, pp. 369–72; *166*

ARCHBALD, W. A., and SMITH, E. J. D. (1953): 'Super-refractories' *in* 'Ceramics: a Symposium' (Green, A. T., and Stewart, G. H., eds.) (British Ceramic Society, Stoke-on-Trent); *254*

ARLETT, P. L. (1965): *Elec. Rev.*, **176**, pp. 712–14; *213*

ARMAN, A. N. and STARR, A. T. (1936): *J. IEE*, **79**, pp. 67–81; *111*

ASTM C, D and E: standards issued by American Society for Testing and Materials, 1916 Race Street, Philadelphia, Pa., USA

ARTBAUER, J., and GRIAČ, J. (1970): *IEEE Trans.*, **EI-5**, pp. 104–12; *63*

ARTS, A. F. M.: see GOLDSCHVARTZ

AUSTIN, J., and JAMES, R. E. (1970): *Conference on dielectric materials etc., Lancaster, pp. 163–6; *114*

AYRES, S.: see LYNCH

BAHDER, G., and GARCIA, F. G. (1971): *IEEE Trans.*, **PAS 90**, pp. 917–25; *148*

BAKER, W. P. (1965a): 'Electrical insulation measurements' (Newnes); *101, 103, 108, 111*

BAKER, W. P. (1965b): AEI Engineering, **5**, pp. 98–102; *103*

BAKER, W. P. (1970): *Conference on dielectric materials etc., Lancaster, pp. 241–4; *219*

BALDWIN, L. V., and HAMILTON, A. N. (1971): *10th electrical insulation conference, Chicago, pp. 87–90; *163*

BALL, E. H., JONES, P. B., SKIPPER, D. J., THELWELL, M. J. and ENDACOTT, J. D. (1972): *CIGRE, paper 21–2, *121, 149*

BALL, J. H.: see NOREN

BARKER, H., and FLACK, T. (1969): *9th electrical insulation conference, Boston, pp. 220–4; *196*

BARNES, C. C. (1966): 'Power cables' (Chapman & Hall); *121, 148*

BARNES, C. C., HILL, E., and SUTTON, C. T. W. (1969): *Proc. IEE*, **116**, pp. 548–60; *147*

BARNETT, M. J.: see HILL, G. J.

BARRIE, I. T., BUCKINGHAM, K. A. and REDDISH, W. (1966): *Proc. IEE*, **113**, pp. 1849–54; *144, 145*

BARRY, C. P. (1970): *BEAMA conference, London, paper *4b*; *119, 121*

BARRY, T. H. (1969): 'Natural resins' *in* 'Paint technology manuals—Pt. 2' (Chapman & Hall, 2nd edn.); *135*

BARTNIKAS, R. (1967): *IEEE Trans.*, **EI-2**, pp. 33–54; *219, 228*

BEACHAM, H. H.: see SEGRO

van BEEK, L. K. H. (1967): 'Dielectric behaviour of heterogeneous systems' *in* 'Progress in dielectrics—Vol. 7' (Birks, J. B., ed.) (Heywood); *56*

BEER, G., GASPARINI, G., OSIMO, F., and ROSSI, F. (1966): *CIGRE, paper 135; *87*

BEHN, R. (1968): *Siemens Zeits.*, *42*, pp. 233–5; *150*

BENNETT, R. G., and CALDERWOOD, J. H. (1971): 'Experimental techniques in dielectric studies' *in* 'Complex permittivity' (Scaife, B. K. P., ed.) (English Universities Press); *101, 104*

BERBERICH, L. J.: see LAFFOON

BILLING, J. W., and MASON, J. H. (1970): *Conference on dielectric materials etc., Lancaster, pp. 93–6; *65*

BILLINGS, M. J., and HUMPHREYS, K. W. (1968): *IEEE Trans.*, **EI-3**, pp. 62–70; *77*

BILLINGS, M. J., WILKINS, R., and WARREN, L. (1968): *IEEE Trans.*, **EI-3**, pp. 33–9; *110*

BILLMEYER, F. W. (1970): 'Text book of polymer science', (2nd edition, Wiley-Interscience); *19, 141, 142, 145*

BINGGELI, J., FROIDEVAUX, J., and KRATZER, R. (1966): CIGRE, paper 110; *83, 85, 87, 217, 218*

BINNS, D. B. (1970): *Proceedings of British Ceramic Society conference, Warwick, pp. 17–22; *258*

BIRKS, J. B. (1960): 'Chlorinated hydrocarbons' and 'Synthetic high polymers' *in* 'Modern dielectric materials' (Birks, J. B., ed.) (Heywood); *184, 186, 221*

BIRLASEKARAN, S., and DARVENIZA, M. (1972): *Conference on dielectric liquids, Dublin, pp. 120–3 and 124–8; *54*

BISHOP, A.: see REDDISH

BLACK, R. M., and CHARLESBY, A. (1960): 'Irradiated polymers' *in* 'Progress in dielectrics—Vol. 2' (Birks, J. B., and Schulman, J. H., eds.) (Heywood); *37*

BLACK, R. M., and REYNOLDS, E. H. (1964): 'Ionization and irradiation effects in high voltage dielectric materials', *Conference on dielectric and insulating materials, London; 37

BLACK, R. M.: see REYNOLDS

BLAISSE B. S.: see GOLDSCHVARTZ

BLANKENBURG, R. C., STEINKE, W. W., and STOPLE, J. (1970): *IEEE Trans.*, **PAS 89**, pp. 304–12; *148*

BLINNE, K., MÄDER, O., and PETER, J. (1961): *Bull. Oerlikon*, 345, pp. 37–60; *131*

BLOK, J., and LE GRAND, D. G. (1969): *J. App. Phys.*, **40**, pp. 288–93; *63*

BLOOR, E. C. (1953): 'Composition and properties of electrical ceramics' *in* 'Ceramics: a Symposium' (Green, A. T., and Stewart, G. H., eds.) (British Ceramic Society, Stoke-on-Trent); *245, 248, 251*

BLOW, C. M. (ed.) (1971): 'Rubber technology and manufacture'; *174*

BOBO, J. C.: see FALLOU, B.

BOGGS, F. W.: see WORKS

BOLTON, B., COOPER, R., and GUPTA, K. G. (1965): *Proc. IEE*, **112**, pp. 1215–20; *64*

BOONE, W., and VERMEER, J. (1970): *Conference on dielectric materials etc., Lancaster, pp. 223–6; *73, 227*

BÖTTCHER, C. J. F. (1952): 'Theory of electric polarisation' (Elsevier); *40*

BOWDLER, G. W.: see RAYNER

BRADWELL, A., COOPER, R. and VARLOW, B. (1971): Proc. IEE, **118**, pp. 247–54; *53, 63*

BRADWELL, A., EDWARDS, D., and HOLT, T. (1970): *Conference on dielectric materials etc., Lancaster, pp. 279–82; *79*

BRAZIER, L. G. (1954): 'Power cables and capacitors' *in* 'The insulation of electrical equipment' (Jackson, W., ed.) (Chapman & Hall); *133*

BREITENSTEIN, A. M., JOHNSTON, D. R. and MAUGHAN, C. V. (1969): *IEEE Trans.*; **PAS 88**, pp. 1389–93; *130*

BRIGNELL, J. E.: see HEWISH

BRIGNELL, J. E.: see RHODES

BROWN, G. P., HAARR, D. T., and METLAY, M. (1973): *IEEE Trans* , **EI-8**, pp. 36–41; *28*

BROWN, J. M.: see GALPERN

BRUINS, P. F.: see ALMOULI

BRYDSON, J. A. (1969): 'Plastic materials' (Iliffe, 2nd edn): *142, 182*

BS NUMBER: British Standard issued by British Standards Institution, 2 Park Street, London, W1

BUCKINGHAM, K. A., and REDDISH, W. (1967): *Proc. IEE*, **114**, pp. 1810–14; *46, 146*

BUCKINGHAM, K. A.: see BARRIE

BUCKINGHAM, K. A.: see REDDISH

BUIST, J. M., and GUDGEON, H. (eds.) (1968): 'Advances in polyurethane technology' (Maclaren & Sons); *202*

BUNAG, M. M. and KOENIG, J. H. (1962): *J. Amer. Ceram. Soc.*, **42**, (9), p. 442; *253*

BURFOOT, J. C. (1967): 'Ferroelectrics' (van Nostrand); *248*

BUTTON, J. C. E., and DAVIES, A. J. (1953): *J. Sci. Instr.*, **30**, pp. 307–10; *210*

C NUMBER: Standard on ceramic and masonry materials issued by ASTM (q.v.)

CALDERWOOD, J. H.: see BENNETT

CALDERWOOD, J. H., see SINGH

CESANA, U. V., MARWICK, I. J., and MCNAMARA, J. H. (1971): Conference on underground distribution, Detroit, pp. 213–12 (New York: Institute of Electrical & Electronics Engineers); *148*

CHADBAND, W. G.: see SINGH

CHARLESBY, A.: see BLACK

CHAROY, M. A., and JOCTEUR, R. F. (1971): *IEEE Trans.*, **PAS 90,** pp. 777–84; *148*

CHEN, D. J., and STALEY, R. W. (1967): *7th electrical insulation conference, Chicago, pp. 5–7; *128*

CHERNEY, E. A., and CROSS, J. D. (1973): *IEEE Trans.*, **EI-8,** pp. 10–16; *60*

CHILDS, D. G., and STANNETT, A. W. (1953): *Proc. IEE,* **100,** pt. IIA, pp. 61–7; *209*

CHURCH, H. F.: see KRASUCKI

CHURCHER, B. G., and DANNATT, C. (1926): *World Power,* **5,** pp. 283–92; *102*

*CIGRE (1968): various authors, papers 12.01, 12.02, 12.04, 12.06, 12.07, 12.09, 12.10, 12.12, 12.15; *114*

CLARK, F. M. (1962): 'Insulating materials for design and engineering practice' (Wiley); *35, 82, 119, 123, 125, 126, 132, 174, 185, 186, 188, 190, 195, 203, 204, 207, 221, 226, 227, 238*

CLOTHIER, N. D., LAWRENCE, R., and DENHAM, J. A. (1970): *BEAMA conference, London, paper 5*b*; *217*

CLULEY, H. J.: see STILL

COLE, R. H. (1961): 'Theories of dielectric polarization and relaxation' *in* 'Progress in dielectrics—Vol. 3' (Birks, J. B., and Hart, J., eds.) (Heywood); *41, 48*

COLEMAN, C. R. (1964): 'Some aspects of askarels as capacitor impregnants', *Conference on dielectric and insulating materials, London; *223, 224*

COLEMAN, C. R. (1970): *BEAMA conference, London, paper 5*d*; *218, 222*

CONSTANDINOU, T. E. (1965*a*): *Insulation,* June, pp. 37–40; *106, 116*

CONSTANDINOU, T. E. (1965*b*): *Proc. IEE,* **112,** pp. 1783–95; *84–85, 106, 116*

CONSTANDINOU, T. E. (1969) *Proc.: IEE,* **116,** pp. 834–46; *71*

COOK, W. R.: see JAFFE, B.

COOPER, R. (1963): 'Electric breakdown of alkali halide crystals' *in* 'Progress in dielectrics—Vol. 5' (Birks, J. B., and Hart, J., eds.) (Heywood); *63*

COOPER, R. (1966): *Brit. J. App. Phys.,* **17,** pp. 149–66; *63*

COOPER, R.: see BOLTON

COOPER, R.: see BRADWELL

COSGROVE, M. F.: see ADLER

COTTRELL, D. W. (1970): *Conference on dielectric materials etc., Lancaster, pp. 322–7; *227*

COUNSELL, J. A. H.: see EDWARDS, D. R.

CROSS, J. D.: see CHERNEY

CURTIS, G. A. (1970): *BEAMA conference, London, paper 4*a*; *150*

D NUMBER: Standards on miscellaneous materials issued by ASTM (q.v.)

DAKIN, T. W., and MALINARIC, P. J. (1960): *AIEE Trans.,* **PAS 79,** pp. 648–53; *113*

DAKIN, T. W.: see WORKS

DALHOFF, W.: see MENGES

DANIEL, V. V (1967): 'Dielectric relaxation' (Academic Press); *21, 22, 41, 48, 50, 56, 59, 60*

DANNATT, C.: see CHURCHER

DARVENIZA, S.: see BIRLASEKARAN

DAVIS, A. J.: see BUTTON

DAVIES, C. L. (1970): *Conference on dielectric materials etc., Lancaster, pp. 287–91; *130*

DAVIES, C. L.: see APPLEBY

DAVIES, I. (1971): *J. App. Chem. & Biotechnol.*, 21, pp. 194–9; *213*

DAVIS, J. H , and JONES, R. T. (1965): *Proc. IEE*, **112,** pp. 1601–6; *201*

DAVIS, J. H., and REES, D. E. W. (1965): *Proc. IEE*, **112,** pp, 1607–13; *179*

DAVIS, R.: see RAYNER

DAY, A. G.: see APPLEBY

DEBYE, P. (1929): 'Polar Molecules' (Chemical Catalog Co.); *41*

DEMATOS, H. V., and HUTZLER, J. R. (1971): Proceedings of the 21st electrical components conference, Washington, DC 10–12 May, pp. 129–36 (New York: Institute of Electrical and Electronics Engineers); *172*

DENHAM, J. A., HAWLEY, R., RICHARDSON, P., DOUGLAS, J. L., and LACOTTA, J. M. (1972): *CIGRE, paper 15–01; *130, 197*

DENHAM, J. A.: see CLOTHIER

DENSLEY, R. J., and SALVAGE, B. (1971): *IEEE Trans.*, **EI-6,** pp. 54–62; *69*

DEUTSCH, K., HOFF, E. A. W., and REDDISH, W. (1951): *J. Polymer Science*, **13,** pp. 565–82; *161*

DEXTER, G. F.: see KOOKOOTSEDES

DEY, P., DRINKWATER, B. J., and PROUD, S. H. R. (1968): Electrical Times, **154,** no. 15, pp. 191–4; *77*

DICKSON, M. R. (1959): ERA report Q/T 152; *87*

DIN NUMBER Deutscher Normenausschuss, Beuth-Vertrieb GmbH, Berlin 30

DONELLY, E. (1971) Electronics & Power, pp. 395–9, *185*

DOO, V. Y., KERR, D. R., and NICHOLS, D. R. (1968): *J. Electrochem. Soc.*, **115,** pp. 61–4; *257*

DORMAN, E. N. (1969): 'Epoxy resins' *in* 'Handbook of fiberglass and advanced composites' (Lubin, G., ed.) (von Nostrand Reinhold); *193*

DÖRNENBURG, E., and GERBER, O. E. (1967): *Brown Boveri Review*, **54,** pp. 104–11; *213*

DOUGLAS, G. C. G.: see APPLEBY

DOUGLAS, J. L.: see DENHAM

DOWLING, D. J. (1960): 'Silicones' *in* 'Modern dielectric materials' (Birks, J. B., ed.) (Heywood); *228*

DOYLE, A. F. (1969): *9th electrical insulation conference, Boston, pp. 85–8; *129*

DOYLE, H. J., and HARRIER, S. C. (1969): 'Phenolics and silicones' *in* 'Handbook of fiberglass and advanced composites' (Lubin, G., ed.) (van Nostrand-Reinhold); *200*

DRINKWATER, B. J.: see DEY

DUBLIN (1972): Various contributors to *Conference on dielectric liquids, Dublin

DUBOIS, J. H. (1966): *Insulation*, Nov., pp. 72–4; *129*

DUNKLEY, J., and SILLARS, R. W. (1953): *Proc. IEE*, **100**, pt. IIA, pp. 73–80, *80, 219*

DURAND, P., and FOURNIÉ, R. (1970): *Conference on dielectric materials etc., Lancaster, pp. 142–5; *59*

E NUMBER: Standards on miscellaneous methods issued by ASTM (q.v.)

EDWARDS, D.: see BRADWELL

EDWARDS, D. R., COUNSELL, J. A. H., GIBBONS, J. A. M., and SCARISBRICK, R. M. (1972): *CIGRE, paper 15–05; *149*

EDWARDS, D. R.: see REYNOLDS

EDWARDS, F. S.: see RYDER

EICHBAUM, B. R. (1972): *Insulation/Circuits*, June–July, pp. 190–200; *245*

ENDACOTT, J. D.: see BALL, E. H.

ERDMAN, D. E., and LAUROESCH, H. C. (1969): *9th electrical insulation conference, Boston, pp. 159–61; *131, 196*

FALLOU, B., and BOBO, J. C. (1970): *BEAMA conference, London, paper 3*a*; *144, 145, 155*

FALLOU, M.: see OUDIN

FARMER, E. (1970): *BEAMA conference, London, paper 1; *131*

FELICI, N. J. (1967): *Direct Current*, **2**, pp. 90–9; *229*

FELICI, N. J., GARTNER, E., and TOBAZÉON, R. (1970): *Conference on dielectric materials etc., Lancaster, pp. 297–304; *229*

FERGESTAD, R., and SCHANCHE, T. (1968): *IEEE Trans.*, **EI-3**, pp. 49–55; *131*

FERRY, J. D. (1970): 'Viscoelastic properties of polymers' (Wiley, 2nd edn.); *19*

FLACK, T.: see BARKER

FLANAGAN, P. E., and JENNY, A. L. (1972): *Insulation/Circuits*, Feb., pp. 35–7; *116*

FLYNN, E. J., KILBOURNE, C. E., and RICHARDSON, C. D. (1958): *AIEE Trans.*, **PAS 77**, pt. III, pp. 358–71; *131, 196*

FORREST, J. S. (1963): *Advanc. Sci.*, Nov., pp. 319–27; *65, Fig. 27 (Plate)*

FORSTER, E. O. (1967): *IEEE Trans.*, **EI-2**, pp. 10–18; *60*

FOURNIÉ, R.: see DURAND

FRANKLIN, E. B. (1965): *Elec. Times*, **147**, pp. 787–91; *88*

FRIEDRICH, K. F.: see PALUMBO

FRÖHLICH, H. (1949): 'Theory of dielectrics' (Clarendon Press); *41*

FROIDEVAUX, J.: see BINGGELI

FUJISAWA, Y., YASUI, T., KAWASAKI, Y., and MATSUMURA, H. (1968): *IEEE Trans.*, **PAS 87**, pp. 1899–907; *148*

GALASSO, F. S. (1969): 'Structure, properties and preparation of perovskite type compounds' (Pergamon, Oxford); *248*

GALPERN, H. N., and BROWN, J. M. (1971): *IEEE Trans.*, **EI-6**, pp. 21–32; *178*

GARCIA, F. F.: see BAHDER

GARTNER, E.: see FELICI

GARTON, C. G. (1941): *J. IEE*, **88**, Pt. 3, pp. 23–40; *60, 229*

GARTON, C. G., and KRASUCKI, Z. (1964): *Proc. Roy. Soc* [A], **280**, pp. 211–26; *73*

GARTON, C. G.: see KRASUCKI

GARTON, C. G.: see MASON

GARTON, C. G.: see STARK

GASPARD, F. (1970): *Conference on dielectric materials etc., Lancaster, pp. 134–7; *229*

GERBER, O. E.: see DÖRNENBURG

GIBBONS, J. A. M., HOWARD, P. R., and SKIPPER, D. J. (1965): *Proc. IEE*, **112**, pp. 89–102; *149*

GIBBONS, J. A. M.: see EDWARDS, D. R.

GIBBONS, J. M., and STANNETT, A. W. (1973): *Proc. IEE*, **120**, pp. 433–9; *149*

GJAJA, N. V.: see KOHN

GOLDENBERG, H. (1962): *IEEE Trans.*, **PAS 80**, pp. 947–54; *30*

GOLDSCHVARTZ, J. M., van STEEG, C., ARTS, A. F. M., and BLAISSE, B. S. (1972): *Conference on dielectric liquids, Dublin, pp. 228–34; *230*

GOSWAMI, B. M., ANGERER, L., and WARD, B. W. (1970): *Conference on dielectric materials etc., Lancaster, pp. 237–44; *217*

GRAY, R., and LEWIS, T. J. (1965): *Br. J. Appl. Phys.*, **16**, pp. 1049–50; *54*

GREEN, W. J., and VERNE, S. (1960): 'Natural and synthetic rubbers' *in* 'Modern dielectric materials' (Birks, J. B., ed.) (Heywood); *174, 177*

GRIAČ, J.: see ARTBAUER

GROSS, B. (1948): *J. Appl. Phys.*, *19*, pp. 257–64; *21, 23*

GROVE, C. S.: see ROSATO

GUDGEON, M.: see BUIST

GUPTA, K. G.: see BOLTON

HAARR, D. T.: see BROWN, G. P.

HACKER, H. (1967): *Radio Mentor*, **33**, pp. 365–7; *166*

HAGUE, B., and FOORD, T. R. (1971): 'A.C. bridge methods' (Pitman, 6th edition); *102*

HALL, H. C., and KELK, E. (1962): 'Paper' *in* 'Modern dielectric materials' (Birks, J. B., ed.) (Heywood); *119*

HAMILTON, A. N.: see BALDWIN

HAMILTON, C. W. (1961): 'Recent developments in cable insulation in US' *in* 'Progress in dielectrics—Vol. 3' (Birks, J. B., and Hart, J., eds.) (Heywood); *147*

HARALDSEN, S., and WINBERG, K. (1968): *CIGRE, paper 12.09; *110*

HARPER, C. A. (1969): 'Materials for electronic packaging' *in* 'Handbook of electronic packaging' (Harper, C. A., ed.) (McGraw-Hill); *153, 155, 196, 240, 253, 254*

HARRIER, S. C.: see DOYLE, H. J.

HARTSHORN, L., PARRY, J. V. L., and RUSHTON, E. (1953): *Proc. IEE*, **100**, pt. IIA, pp. 23–37; *43*

HARTSHORN, L., and WARD, A. (1936): *J. IEE*, **79**, pp. 597–609; *103, 105*

HARVEY, A. F. (1963): 'Microwave engineering' (Academic Press); *104*

HAWLEY, R.: see DENHAM

HEIDLER, G. R. (1969): 'Depositions for microelectronics' *in* 'Handbook of electronic packaging' (Harper, C. A., ed.) (McGraw-Hill); *240, 241*

HETHERINGTON, W., and KEIL, C. (1966): *CIGRE, paper 142, app. 2; *88*

HEWISH, T. R., and BRIGNELL, J. E. (1972): *J. Phys. D*, **5**, pp. 747–52; *52, 54*

HIGGIN, R. M., and HIRSCH, J. (1970): *Conference on dielectric materials etc., Lancaster, pp. 210–13; *50*

HIGHAM, J. B.: see WATSON

HILL, C. F.: see LAFFOON

HILL, E.: see BARNES

K*

HILL, G. J., STRUDWICK, P., and BARNETT, M. J. (1970): *British Ceramic Society conference, Warwick, pp. 263–70; *254*

V. HIPPEL, A. R. (1954): 'Dielectric materials and applications' (MIT, John Wiley and Chapman & Hall); *101, 104, 223, 225, 228, 233, 253*

HIRSCH, J., see HIGGIN

HOFF, E. A. W., see DEUTSCH

HOGG, W. K., and WALLEY, C. A. (1970): *Proc. IEE*, **117,** pp. 261–8; *66*

HOLT, T.: see BRADWELL

HOPKINSON, J. (1876): *Phil. Trans. Roy. Soc.*, **166,** pp. 489–94; *47*

HOPKINSON, J. (1897): *Phil. Trans. Roy. Soc.*, **189A,** pp. 109–34; *48*

HORNSBY, E. A., IRVING, R., and PATTERSON, E. A. (1965a): *Proc. IEE*, **112,** pp. 586–9; *211*

HORNSBY, E. A., IRVING, R., and PATTERSON, E. A. (1965b): *Proc IEE*, **112,** pp. 590–6; *213*

HOUSE, H.: see POLLARD

HOUSE, H.: see TAYLOR, R. J.

HOWARD, C. P.: see APPLEBY

HOWARD, P. R. (1951): *Proc. IEE*, **98,** Pt. 2, pp. 365–70; *66, 70*

HOWARD, P. R.: see GIBBONS, J. A. M.

HUMPHREYS, K. W.: see BILLINGS

HUTZLER, J. R.: see DEMATOS

HYDE, P. J. (1970): *Proc. IEE*, **117,** pp. 1891–901; *48, 104*

HYDE, P. J.: see REDDISH

IEC NUMBER: Publication issued by the International Electrotechnical Commission, Geneva, Switzerland

IEEE NUMBER: Standards publication issued by the Institute of Electrical and Electronics Engineers, 345 East 47th Street, New York, NY, USA

IKEMOTO, N.: see ISOGAI

ILLERS, F., and KUHMANN, H. (1969): *Elektrotech. Z.*, **21B,** pp. 555–8, *148*

INQUE, T.: see ISOGAI

INSULATION/CIRCUITS ENCYCLOPEDIA (Annual), Lake Publishing Corpn., Lybertyville, Illinois, USA

IP NUMBER: Standard from 'Standard methods of testing petroleum and its products', issued by Institute of Petroleum, 61 New Cavendish Street, London W1

IRVING, R.: see HORNSBY

ISARD, J. O. (1962): *Proc. IEE*, **109B,** Suppl. 22, pp. 440–7; *59*

ISOBE, S.: see MATSUNOBU

ISOGAI, T., IKEMOTO, N., and INQUE, T. (1971): *10th electrical insulation conference, Chicago, pp. 313–8; *197*

ISO/R NUMBER: Recommendation issued by International Organisation for Standardization, through British Standards Institution

JAFFE, B., COOK, W. R., and JAFFE, H. (1971): 'Piezoelectric ceramics' (Academic Press); *248*

JAFFE, H.: see JAFFE, B.

JAMES, R. E.: see AUSTIN

JEFFERIES, M. J., and MATHES, K. N. (1970): *IEEE Trans.*, **EI-5,** pp. 83–91; *73, 230, 231*

JENNY, A. L.: see FLANAGAN

JOCTEUR, R. F.: see CHAROY

JOHNSTON, D. R.: see BREITENSTEIN

JOLLEY, H. E. W. (1965): *Proc. IEE*, **112**, pp. 1061–70; *70, 185*

JONES, E. (1952): 'Electrical machines' *in* 'The insulation of electrical equipment' (Jackson, W., ed.) (Chapman & Hall); *131*

JONES, P. B.: see BALL, E. H.

JONES, R. T.: see DAVIS, J. H.

KANISKIN, V. A., TRAN, H. T., and SHEN, M. (1972): *Insulation/Circuits*, May, pp. 37–41; *150*

KAUFMAN, R. B., SHIMANSKI, E. J., and MACFADYEN, K. W. (1965): *AIEE Trans.*, **74**, Pt. I, pp. 312–8; *214, 216*

KAWAGUCHI, Y., and YANABU, S. (1969): *IEEE Trans.*, **PAS 88**, pp. 1187–94; *114*

KAWASAKI, T.: see FUJISAWA

KEENAN, T. F. (1972): *Conference on dielectric liquids, Dublin, pp. 235–8; *230*

KEIL, C.: see HETHERINGTON

KELK, E., and WILSON, I. O. (1965): *Proc. IEE*, **112**, pp. 602–12; *119*

KELK, E.: see HALL

Kendall, P. G.: see MOLE

KERR, D. R.: see DOO

KETTERER, R. J. (1964): *Insulation*, Aug., pp. 24–32; *128, 129*

KETTLER, H., and LANGE, H. (1956): *Siemens Zeits.*, **30**, pp. 319–26; *87*

KILBOURNE, C. E.: see FLYNN

KING, A., and WENTWORTH, V. H. (1954): 'Raw materials for electric cables' (Ernest Benn); *119*

KINGSTON, R. G.: see STANNETT

KITCHIN, D. W., and PRATT, O. S. (1958): *AIEE Trans.*, Pt. III, **PAS 77**, pp. 180–6; *63*

KOENIG, J. H.: see BUNAG

KOGAN, P. (1963): *Proc. IEE*, **110**, pp. 2257–66; *83*

KOHL, W. H. (1964): *Vacuum*, **14**, pp. 333–54; *258*

KOHN, L. S., GJAJA, N. V., and URBAN, R. C. (1971): *10th electrical insulation conference, Chicago, pp. 257–9

KOK, J. A. (1961): 'Electrical breakdown of insulating liquids' (Philips, Eindhoven); *217*

KOOKOOTSEDES, G. J., and DEXTER, J. F. (1968): 'Silicone plastics for electrical insulation' *in* 'Plastics for electrical insulation' (Bruins, P. F., ed.) (Wiley Interscience); *179*

KRASUCKI, Z. (1962) *Proc. IEE*, **109B**, Suppl. 22, pp. 435–9; *74*

KRASUCKI, Z. (1966); *Proc. Roy. Soc.* (A), **294**, pp. 393–404; *74*

KRASUCKI, Z. (1967a) ERA report 5212; *54*

KRASUCKI, Z. (1967b) ERA report 5197; *150*

KRASUCKI, Z. (1970) *BEAMA conference, London, paper 4; *151*

KRASUCKI, Z. (1972) *Conference on dielectric liquids, Dublin, pp. 129–39; *54, 75*

KRASUCKI, Z., CHURCH, H. F., and GARTON, C. G. (1960) *J. Electrochem. Soc.*, **107**, pp. 598–602; *83*

KRASUCKI, Z. see GARTON

KRATZER, R.: see BINGGELI

KREUGER, F. H. (1966): *CIGRE, paper 209, apps. 1 and 2; *114*

KUFFEL, E.: see ALLAN

KUHMANN, H.: see ILLERS

KURTZ, M. (1971): *IEEE Trans.*, **EI-6**, pp. 76–81; *110*

KUSAY, R. G. P. (1971): *Elec. Review*, **188**, 17, pp. 551–6; *87*

KUTI, A. J.: see PALUMBO

KUZAWINSKI, J. G., and WOLFF, G. M. (1970): *IEEE Trans.*, **PAS 89**, pp. 1022–30; *131*

LACOTTA, J. M.: see DENHAM

LAFFOON, C. M., HILL, C. F., MOSES, G. L., and BERBERICH, L. J. (1951): *AIEE Trans.*, **70**, pp. 721–30; *131*

LAMBETH, P. J.: see STANNETT

LAMPE, W. (1969): *Archiv. fur Elektrotech.*, **53**, pp. 121–32; *86*

LANGE, H.: see KETTLER

LAUROESCH, H. C.: see ERDMAN

LAWRENCE, R.: see CLOTHIER

LE GRAND, D. G.: see BLOK

LEWIS, T. J. (1959): 'The electric strength and high-field conductivity of dielectric liquids' *in* 'Progress in dielectrics—Vol. 1' (Birks, J. B., and Schulman, J. H., eds.) (Heywood); *52, 73, 230*

LEWIS, T. J.: see GRAY

LEWIS, T. J.: see SWAN

LEWIS, T. J.: see TAYLOR, D. M.

LOVENGUTH, R. F. (1972): *Insulation/Circuits*, Sept., pp. 35–8; *164*

LUBIN, G. (ED.) (1969): 'Handbook of fiberglass and advanced composites' (van Nostrand Reinhold); *181, 195, 201*

LYNCH, A. C. (1966): *Proc. IEE*, **112**, pp. 426–31; *103, 105*

LYNCH, A. C. (1971): *Proc. IEE*, **118**, pp. 244–6; *50*

LYNCH, A. C., and AYRES, S. (1972): *Proc. IEE*, **119**, pp. 767–70; *104*

MCCLAIN, R. D.: see MERCIER

MACDONALD, D. (1954): 'Power transformers' *in* 'The insulation of electrical equipment' (Jackson, W., ed.) (Chapman & Hall); *122*

MCDONALD, J. L. (1969): *9th electrical insulation conference, Boston, pp. 88–91; *129*

MACFADYEN, K. W.: see KAUFMAN

MCKEOWN, J. J. (1965): *Proc. IEE*, **112**, pp. 824–30, *62*

MCKEOWN, J. J., and OLYPHANT, M. (1965): *6th electrical insulation conference, New York, pp. 125–8; *129*

MCMAHON, E. J., and PERKINS, J. R. (1963): *IEEE Trans.*, **PAS 69**, pp. 1128–36; *65*

MCMAHON, E. J., and PERKINS, J. R. (1964): *IEEE Trans.*, **PAS 82**, pp. 1253–61; *64,*

MCMILLAN, P. W. (1964): 'Glass-ceramics' (Academic Press); *243, 244*

MCMILLAN, R. E., and MISRA, R. P. (1970): *IEEE Trans.*, **EI-5**, pp. 10–18; *257*

MCNEALL, P. I.: see SKIPPER

MÄDER, O.: see BLINNE

MAHON, L. P. (1969): *IEEE Trans.*, **PAS 88**, pp. 258–66; *30, 97, 100*

MALINARIC, P. J.: see DAKIN

MARLOW, J. H. (1970): *BEAMA conference, London, paper 5c; *71, 122, 185*

MARTIN, S. R. W. (1972): 'Oleoresinous varnishes', 'Alkyd resins' and 'Amine resins' *in* 'Paint technology manuals part 3' (Chapman & Hall, 2nd edn.); *134, 186, 187*

MASON, J. H. (1959): 'Dielectric breakdown in solid insulation' *in* 'Progress in dielectrics—Vol. 1' (Birks, J. B., and Schulman, J. H., eds.) (Heywood); *62, 66*

MASON, J. H. (1960): *Proc. IEE*, **107A**, pp. 551–68; *66*

MASON, J. H. (1965): *Proc. IEE*, **112**, pp. 1407–23; *66, 111*

MASON, J. H.: see BILLING

MASON, J. H., and GARTON, C. G. (1959): 'Insulation for small transformers', Electrical Research Association, Leatherhead; *215, 216*

MASSEY, L. (1952): *J. Inst. Petroleum*, **38**, pp. 164–71, 281–97, 361–93; *209*

MASSEY, L., and ROMNEY, J. (1965): ERA report 5090; *211*

MASSEY, L., and WILSON, A. C. M. (1958): *J. Inst. Petroleum*, **44**, pp. 336–56; *210*

MATHES, K. N. (1967): *IEEE Trans.*, **EI-2**, pp. 24–32; *230, 231*

MATHES, K. N. (1969): *IEEE Trans.*, **EI-4**, pp. 2–7; *82*

MATHES, K. N.: see JEFFERIES

MATSUMURA, H.: see FUJISAWA

MATSUNOBU, K., ISOBE, S., and MUKAI, J. (1972): *IEEE Trans.*, **EI-7**, pp. 132–39; *131*

MATUSZEWSKI, T., TERLECKI, J., and SULOCKI, J. (1972): *Conference on dielectric liquids, Dublin, pp. 189–94; *54*

MAUGHAN, C. V.: see BREITENSTEIN

MEADOWCROFT, D. B. (1970): *British Ceramic Society conference, Warwick, pp. 7–16; *248*

MEAKINS, R. J. (1961): 'Mechanisms of dielectric absorption in solids' *in* 'Progress in dielectrics—Vol. 3'; (Birks, J. B., and Hart, J., eds.) (Heywood); *43*

MEATS, R. J. (1972): *Proc. IEE*, **119**, pp. 760–6; *73, 231*

MELCHIORE, J. J., and MILLS, I. W. (1967): *IEEE Trans.*, **EI-2**, pp. 150–5; *210, 226*

MENGES, G. and DALHOFF, W. (1972): Quoted in *Design and Engineering*, 10 May 1972; *141*

MERCIER, G., and MCCLAIN, R. D. (1968): *Insulation*, May, pp. 37–8; *150*

MERTENS, W., MEYER, H., and WICHMAN, A. (1967): *7th electrical insulation conference, Chicago, pp. 103–6; *131*

METLAY, M.: see BROWN, G. P.

METTES, D. G. (1969): 'Glass fibers' *in* 'Handbook of fiberglass and advanced composites' (Lubin, G., ed.) (van Nostrand and Reinhold); *241*

MEYER, H.: see MERTENS

MILLS, I. W.: see MELCHIORE

MILLS, M. R. (1952): 'An introduction to drying oil technology' (Pergamon, Oxford); *134*

MISRA, R. P.: see MCMILLAN, R. E.

MITCHELL, J., and SMITH, D. M. (1954): 'Aquametry' (Interscience); *116*

MITRA, G., SAKR, M. M., and SALVAGE, B. (1965): *Proc. IEE*, **112**, pp. 1056–60; *113*

MIYASHITA, T. (1971): *IEEE Trans.*, **EI-6**, pp. 129–35; *81*

MOLE, G. (1953): *Proc. IEE*, **100**, Pt. IIA, pp. 276–83; *105*

MOLE, G. (1962): ERA report V/T 149; *111*

MOLE, G. (1967): ERA report 5186; *111*

MOLE, G., PARROTT, P. G., and KENDALL, P. G. (1969): *Proc. IEE*, **116**, pp. 847–56; *114*

MOREY, G. W. (1954): 'The properties of glass' (Reinhold, 2nd edn.); *238*

MOSES, G. L.: see LAFFOON

MUKAI, J.: see MATSUNOBU

MURTY, N. N.: see MASSEY

NELSON, J. K., SALVAGE, B., and SHARPLEY, W. A. (1971): *Proc. IEE*, **118**, pp. 388–93; *219*

NELSON, W. B. (1971/72): *IEEE Trans.*, **EI-6**, pp. 165–81, and **EI-7**, pp. 36–55; *30, 99*

NEMA: National Electrical Manufacturers Association, 155 East 44th Street, New York, NY, USA:

> LI 1: Publication on industrial laminates and thermosetting products
>
> LI 3: Publication on high temperature properties of industrial Thermosetting laminates

NICHOLS, D. R.: see DOO

NOREN, J. R., and BALL, J. H. (1969): *9th electrical insulation conference, Boston, pp. 1–3; *129*

NORRIS, E. T. (1963): *Proc. IEE*, **110**, pp. 428–40; *211, 217*

NOSHAY, A. (1973): *Insulation/Circuits*, May 1973, pp. 33–8; *196*

OBURGER, W. (1957): 'Die Isolierstoffe der Elektrotechnik' (Springer); *123, 185, 186*

OGORKIEWICZ, R. M. (ed.) (1970a): 'Engineering properties of thermoplastics' (Wiley); *23, 24, 92, 146, 154, 157, 167*

OGORKIEWICZ, R. M. (1970b): 'Mechanical behaviour of fibre composites' *in* 'Glass reinforced plastics' (Parkyn, B., ed.) (Iliffe); *23, 92, 191*

OLYPHANT, M. (1963): *IEEE Trans.*, **PAS 82**, pp. 1106–12; *64*

OLYPHANT, M.: see MCKEOWN

O'TOOLE, J. L.: see RIDDELL

OUDIN, J. M., FALLOU, M., and THÉVENON, H. (1967): *IEEE Trans.*, **PAS 86**, pp. 304–11; *121*

OUWERKERKE, A. C.: see GOLDSCHVARTZ

PALUMBO, A. J., FRIEDRICH, K. F., KUTI, A. J., and WOODS, E. E. (1967): *7th electrical insulation conference, Chicago, pp. 294–9; *77*

PARKMAN, N. (1959): 'Physical properties of polymers', SCI monograph 5, pp. 95–120 (Society of Chemical Industry, London); *69*

PARKMAN, N. (1961): *Proc. IEE*, **109**, Pt. B, Suppl. 22, pp. 448–53; *109*

PARKYN, B. (ED.) (1970): 'Glass reinforced plastics' (Butterworth); *181, 191, 195, 201*

PARR, D. J., and SCARISBRICK, R. M. (1965): *Proc. IEE*, **112**, pp. 1625–32; *77*

PARR, D. J.: see STANNETT

PARRISS, W. H. (1971): *J. Sci. & Technol.*, **38**, pp. 157–66; *131*

PARRISS, W. H.: see APPLEBY

PARROTT, P. G.: see MOLE

PARRY, J. V. L.: see HARTSHORN

PARTRIDGE, J. H. (1952): 'Glass to metal seals' (Society of Glass Technology); *240*

PATTERSON, E. A.: see HORNSBY

PAYNE, A. R. (1959): 'Physical properties of polymers' SCI monograph 5, pp. 273–89 (Society of Chemical Industry, London); *47, 175*

PERKINS, J. R.: see MCMAHON

PETER, J.: see BLINNE

PILPEL, N. (1968): *Insulation*, May, pp. 63–9; *226, 228*

PILPEL, N., and REYNOLDS, E. H. (1960): 'Hydrocarbon insulating oils' *in* 'Modern dielectric materials' (Birk, J. B., ed.) (Heywood); *207*

PIPER, J. D. (1946): *AIEE Trans.*, **65**, pp. 791–7; *83, 84*

POLLARD, A. F., and HOUSE, H. (1972): *Conference on dielectric liquids, Dublin, pp. 23–6; *60*

POPPER, P. (1959): 'Non-oxide ceramic dielectrics' *in* 'Progress in dielectrics—Vol. 1' (Birks, J. B., and Schulman, J. H., eds.) (Heywood); *256*

POPPER, P. (1960): 'Ceramics' *in* 'Modern dielectric materials' (Birks, J. B., ed.) (Heywood); *245, 248*

POTTER, W. G. (1970): 'Epoxide resins' (Iliffe); *193*

PRATT, O. S.: see KITCHIN

PROUD, S. H. R.: see DEY

R NUMBER: see ISO R NUMBER

RAYNER, E. H., STANDRING, W. G., DAVIS, R., and BOWDLER, G. W. (1930): *J. IEE*, **68**, pp. 1132–42; *102*

REDDISH, W. (1950): *Trans. Faraday Soc.*, **46**, pp. 459–75; *44, 45, Figs. 21, 22 (Plates), 82*

REDDISH, W. (1962): *Pure and Applied Chemistry*, **5**, pp. 723–42; *47, 51, 159, 161*

REDDISH, W. (1966): *J. Polymer Science C*, 14, pp. 123–37; *157, 158, 160*

REDDISH, W., BISHOP, A., BUCKINGHAM, K. A., and HYDE, P. J. (1971): *Proc. IEE*, **118**, pp. 255–65; *104*

REDDISH, W., and TAPLEY, J. G. (1970): *BEAMA conference, London, paper 4c; *150*

REDDISH, W.: see BARRIE

REDDISH, W.: see BUCKINGHAM

REDDISH, W.: see DEUTSCH

REES, D. E. W.: see DAVIS, J. H.

REESE, E. (1970): *BEAMA conference, London, paper 7d; *166*

REMBOLD, H. (1964): *Bull. Assoc. Suisse des Electriciens*, **55**, pp. 1025–30; *129, 131, 178*

REYNOLDS, E. H., and BLACK, R. M. (1972): *Proc. IEE*, **119**, pp. 497–504; *213*

REYNOLDS, E. H., and EDWARDS, D. R. (1970): IEE/ERA Conference on Distribution, Edinburgh, 20–22 Oct. 1970, pp. 318–24; *147, 173* (London: Institution of Electrical Engineers)

REYNOLDS, E. H.: see BLACK

REYNOLDS, E. H.: see PILPEL

RHODES, G. M., and BRIGNELL, J. E. (1972) *Conference on dielectric liquids, Dublin, pp. 116–9; *54*

RICE, H. L. (1971): *10th electrical insulation conference, Chicago, pp. 249–53; *163*

RICHARDSON, C. D.: see FLYNN

RICHARDSON, P.: see DENHAM

RICKETTS, C. E. (1970): *BEAMA conference, London, paper 6e; *249, 250*

RIDDELL, M. N., and O'TOOLE, J. L. (1968): *Modern Plastics*, May, p. 150; *92*

RIDDLESTONE, H. G. (1953): *Proc. IEE*, **100**, Pt. IIA, pp. 159–62; *62*

ROFF, W. J., and SCOTT, J. R. (1971) 'Fibres, films, plastics and rubbers' (Butterworth); *141, 174, 193, 199, 202*

ROGERS, D. A. (1967): *7th electrical insulation conference, Chicago, pp. 100–2, *131, 191*

ROLLINSON, W. (1957): *Metropolitan -Vickers Gazette*, **28**, pp. 286–9; *67*

ROMNEY, J.: see MASSEY

ROSATO, D. V., and GROVE, C. S. (1964): 'Filament winding' (Interscience); *242*

RUBIN, M. (1969): 'Polyester resins' *in* 'Handbook of fiberglass and advanced composites' (Lubin, G., ed.) (Van Nostrand Reinhold); *191*

RUSHTON, E.: see HARTSHORN

RYDER, D. M., and EDWARDS, F. S. (1964): 'The breakdown strength of transformer oil at power frequency', *Conference on dielectric and insulating materials, London; *217*

SAKR, M. M.: see MITRA

SALENSKY, G. (1972): *Insulation/Circuits*, May, pp. 19–25; *197, 198*

SALVAGE, B. (1968): Colloquium on charge movement and discharge in solid and liquid dielectrics, 16 May, pp. 15–17 (London: Institution of Electrical Engineers); *70*

SALVAGE, B.: see DENSLEY

SALVAGE, B.: see MITRA

SALVAGE, B.: see NELSON, J. K.

SAUVIAT, M., and TOBAZÉON, R. (1970): *Conference on dielectric materials etc., Lancaster, pp. 227–8; *53*

SCARISBRICK, R. M.: see EDWARDS, D. R.

SCARISBRICK, R. M.: see PARR

SCARISBRICK, R. M.: see STANNETT

SCHANCHE, T.: see FERGESTAD

SCOTT, J. R.: see ROFF

SEGRO, N. R., and BEACHAM, H. H. (1969): *9th electrical insulation conference, Boston, pp. 98–100; *199*

SHARBAUGH, A. H., and WATSON, P. K. (1962): 'Conduction and breakdown in liquid dielectrics', pp. 199–248 of 'Progress in dielectrics— Vol. 4' (Birks, J. B., and Hart, J., eds.) (Heywood); *73, 230*

SHARPLEY, W. A.: see NELSON, J. K.

SHEARING, H. J. (1972): 'Polyurethane surface coatings' *in* 'Paint technology manuals—Pt. 3' (Chapman & Hall); *202*

SHEN, M.: see KANISKIN

SHIBLEY, A. M. (1969): 'Filament winding' *in* 'Handbook of fiberglass and advanced composites' (Lubin, G., ed.) (Van Nostrand Reinhold); *242*

SHIMANSKI, E. J.: see KAUFMAN

SHROFF, D. H., and WILSON, A. C. M. (1967): *Proc. IEE*, **114**, pp. 817–23; *211*

SILLARS, R. W. (1937): *J. IEE*, **80**, pp. 378–94; *57, 81*

SILLARS, R. W.: see DUNKLEY

SIMONS, A.: see VINCENT, R. S.

SIMONS, J. S. (1964): 'The measurement of integrated discharge energy using a dielectric loss analyser', *Conference on dielectric and insulating materials, London; *112*

SINGH, B., CHADBAND, W. G., SMITH, C. W., and CALDERWOOD, J. H. (1972): *J. Phys. D*, **5**, pp. 1457–64, *74*

SKIPPER, D. J., and MCNEAL, P. I. (1965): *Proc. IEE*, **112**, pp. 103–8; *149*

SKIPPER, D. J.: see BALL, E. H.

SKIPPER, D. J.: see GIBBONS, J. A. M.

SMITH, C. M. (1969): 'Bitumens and pitches' *in* 'Paint technology manuals —Pt. 2' (Chapman & Hall, 2nd edn.); *132*

SMITH, C. W.: see SINGH

SMITH, D. M.: see MITCHELL

SMITH, E. J. D.: see ARCHBALD

SMYSER, R. P. (1969): *9th electrical insulation conference, Boston, pp. 141–3; *66, 129*

SMYTH, C. P. (1955): 'Dielectric behaviour and structure' (McGraw-Hill); *40, 48*

SNADOW, R. (1954): 'Classification and review of insulating materials' *in* 'The insulation of electrical equipment' (Jackson, W., ed.) (Chapman & Hall); *32*

SPITZER, F. (1970): *Conference on dielectric materials etc., Lancaster, pp. 333–5; *73*

STALEY, R. W.: see CHEN

STANDRING, W. G.: see RAYNER

STANNETT, A. W., LAMBETH, P. J., PARR, D. J., SCARISBRICK, R. M., WILSON, A., and KINGSTON, R. G. (1969): *Proc. IEE*, **116**, pp. 261–72; *77*

STANNETT, A. W.: see CHILDS

STANNETT, A. W.: see GIBBONS, J. M.

STANWORTH, J. E. (1950): 'Physical properties of glass' (Clarendon); *235, 237, 238*

STARK, K. H., and GARTON, C. G. (1955): *Nature*, **176**, p. 1225; *63*

STARR, A. T.: see ARMAN

van STEEG, C.: see BOLDSCHVARTZ

STEINKE, W. W.: see BLANKENBURG

STELMAK, J. P.: see SZEDON

STEVELS, J. M. (1957): 'Electrical properties of glass' *in* 'Encyclopedia of physics—Vol. 20' (Springer); *59, 239*

STILL, J. E., and CLULEY, H. J. (1972): *Analyst*, **97** (1150), pp. 1–16, *116*

STOEVER, H. J. (1941): 'Applied heat transmission'; *35* (McGraw-Hill)

STOLPE, J.: see BLANKENBURG

STRUDWICK, P.: see HILL, G. J.

SULOCKI, J.: see MATUSZEWSKI

SUTTON, P. M. (1960): 'The dielectric properties of glass' *in* 'Progress in dielectrics—Vol. 2' (Birks, J. B., and Schulman, J. H., eds.) (Heywood); *51, 59, 238*

SUTTON, C. T. W.: see BARNES

SWAN, D. W., and LEWIS, T. J. (1960): *J. Electrochem. Soc.*, **107**, pp. 180–5; *230*

SZEDON, J. R., and STELMAK, J. P. (1970): *IEEE Trans.*, **EI-5**, pp. 3–9; *257*

TAPLEY, J. G.: see REDDISH

TAYLOR, D. M., and LEWIS, T. J. (1971): *J. Phys. D*, **4**, pp. 1346–57; *53*

TAYLOR, H. E. (1957): *J. Soc. Glass Technol.*, **41**, pp. 350T–82T; *239*

TAYLOR, H. E. (1960): 'Glass' *in* 'Modern dielectric materials' (Birks, J. B., ed.) (Heywood); *238*

TAYLOR, R. J., and HOUSE, H. (1972): *Conference on dielectric liquids, Dublin, pp. 1–4; *52*

TERLECKI, J.: see MATUSZEWSKI

THELWELL, M. J.: see BALL, E. H.

THÉVENON, H.: see OUDIN

THOMPSON, C. N.: see MASSEY

TOBAZÉON, R.: see FELICI

TOBAZÉON, R.: see SAUVIAT

TOOP, D. J. (1971): *IEEE Trans.*, **EI-6**, pp. 2–14; *28*

TRAN, H. T.: see KANISKIN

URBAN, R. C.: see KOHN

VAHLSTROM, W. JR. (1971): Conference on underground distribution, Detroit, pp. 222–9 (New York: Institute of Electrical and Electronics Engineers); *63*

VARLOW, B.: see BRADWELL

VDE NUMBER: Recommendation, regulation, specification issued by Verband Deutscher Elektrotechniker, VDE-Verlag GmbH, Berlin 12

VERMEER, J.: see BOONE

VERNE, S.: see GREENE

VINCENT, R. S., and SIMONS, A. (1940): *Proc. Phys. Soc.*, **52**, pp. 487–500; *116*

VINCETT, P. S. (1969): *Br. J. Appl. Phys.*, series 2, Vol. 2, pp. 699–710; *144, 147, 155*

WADDINGTON, F. B. (1959): 'Laboratory practice 8', pp. 275–8; *116, 209*

WADDINGTON, F. B. (1962): *AEI Engineering*, **2**, pp. 24–9; *37*

WADDINGTON, F. B., and ALLAN, D. J. (1969): *Elec. Rev.*, 184, pp. 751–5; *213*

WALLEY, C. A.: see HOGG

WARD, A.: see HARTSHORN

WARD, B. W.: see GOSWAMI

WARING, L. A. R. (1970): 'Reinforcement' *in* 'Glass-reinforced plastics' (Parkyn, B., ed.) (Butterworth); *241, 242*

WARREN, H. (1931): 'Electrical insulating materials' (Ernest Benn); *118, 123, 132, 134*

WARREN, L.: see BILLINGS

WATSON, P. K., and HIGHAM, J. B. (1953): *Proc. IEE*, **100**, Pt. IIA, pp. 168–74; *217*

WATSON, P. K.: see SHARBAUGH

WAYE, B. E. (1967): 'Introduction to technical ceramics' (Maclaren & Sons); *245, 248, 254, 257*

WENTWORTH, V. H.: see KING

WHITEHEAD, S. (1951): 'Dielectric breakdown of solids' (Clarendon); *70, 71, 72*

WHITEHEAD, S. (1954): 'Dielectric breakdown' *in* 'The insulation of electrical equipment' (Jackson, W., ed.) (Chapman & Hall); *70*

WICHMAN, A.: see MERTENS

WILKINS, R.: see BILLINGS

WILSON, A.: see STANNETT

WILSON, A. C. M. (1965): *Proc. IEE*, **112**, pp. 617–32; *208, 210*

WILSON, A. C. M.: see MASSEY

WILSON, A. C. M.: see SHROFF

WILSON, I. O.: see KELK
WINDBERG, K.: see HARALDSEN
WOLFF, G. M.: see KUZAWINSKI
WOODS, E. E.: see PALUMBO
WORKS, C. N., DAKIN, T. W., and BOGGS, F. W. (1944): *AIEE Trans.*, *63*, pp. 1092–8 and 1452–4; *104*
WYCKOFF, R. W. G. (1948, 1957): 'Crystal structures—Vols. 1 and 3' (Wiley); *Figs. 1, 2 and 36 (Plates)*
WYLLIE, G. (1960): 'Theory of polarization and absorption in dielectrics' *in* 'Progress in dielectrics—Vol. 2; *40* (Birks, J. B., and Schulman, J. H., eds.) (Heywood)
YANABU, S.: see KAWAGUCHI
YASUI, T.: see FUJISAWA

*Conference details

1970 BEAMA conference, London
      BEAMA electrical insulation conference, London 8–10th April (London: British Electrical & Allied Manufacturers' Association)
1970 British Ceramics Society Conference, Warwick
      Proceedings of the British Ceramic Society conference on electrical and magnetic materials (2), Warwick 9–10th Sept. (Stoke-on-Trent; British Ceramic Society)
1970 Conference on dielectric materials etc., Lancaster
      Proceedings of the conference on dielectric materials, measurements and applications, Lancaster 20–24th July (London: Institution of Electrical Engineers)
1964 Conference on dielectric and insulating materials, London. (London: Institution of Electrical Engineers)
1965 6th electrical insulation conference, New York 13–16th Sept.
1967 7th electrical insulation conference, Chicago 15–19th Oct.
1969 9th electrical insulation conference. Boston 8–11th Sept.
1971 10th electrical insulation conference, Chicago 20–23rd Sept.
      (New York: Institute of Electrical and Electronics Engineers)
CIGRE
      International conferences on large high-tension electric systems; 1966 8–18th June, 1968, 10–20th June, 1972, 28th Aug.–6th Sept. (Paris: Conférence Internationale des Grandes Résaux Electriques)
1972 Conference on dielectric liquids, Dublin
      Fourth international conference on conduction and breakdown in dielectric liquids, Dublin 25–27th July (Dublin: Typografia Hiberniae)

Index